AF303157

Novel Planar Metamaterial Resonators
-Design and Study-

Von der Fakultät für Elektrotechnik, Informationstechnik, Physik
der Technischen Universität Carolo-Wilhelmina zu Braunschweig

zur Erlangung der Würde

eines Doktor-Ingenieurs (Dr.-Ing.)

genehmigte Dissertation

von M.Sc. Ibraheem Abdalwahhab Ibraheem Al-Naib
aus Mosul-Nineveh, Irak

eingereicht am: 4. Nov. 2009

mündliche Prüfung am: 11. Dez. 2009

Berichterstatter: Prof. Dr. rer. nat. Martin Koch

Prof. Dr.-Ing. Ludwig Wetenkamp

Prof. Dr.-Ing. Jörg Schöbel

2010

Bibliografische Information der Deutschen Nationalbibliothek

Die Deutsche Nationalbibliothek verzeichnet diese Publikation in der Deutschen Nationalbibliografie; detaillierte bibliografische Daten sind im Internet über http://dnb.d-nb.de abrufbar.

1. Aufl. - Göttingen : Cuvillier, 2010
  Zugl.: (TU) Braunschweig, Univ., Diss., 2009

978-3-86955-234-7

Printed with the support of DAAD (Deutscher Akademischer Austauschdienst)

© CUVILLIER VERLAG, Göttingen 2010
  Nonnenstieg 8, 37075 Göttingen
  Telefon: 0551-54724-0
  Telefax: 0551-54724-21
  www.cuvillier.de

1. Auflage, 2010
Gedruckt auf säurefreiem Papier

978-3-86955-234-7

To my family

"Geht nicht, gibt's nicht."

"Wer's abwarten kann, kommt endlich dran."

Acknowledgements

I would like to express my deep gratitude and thanks to my doctoral father, Professor Martin Koch for his supervision, guidance and invaluable support over the past four years. His generosity, faith and encouragement have always persevered through the years. Thanks are due to Professor Jörg Schöbel for his help and discussions, particularly in the beginning of my work. The advices of Professor Wael Adi are also acknowledged.

I have been extremely fortunate to get a chance to work with the terahertz group and be in a direct contact with two other groups, namely the microwave group and the mobile communication group, all at TU Braunschweig. Working at this interdisciplinary place was certainly an excellent experience. I would like to thank current and past members of the terahertz communication lab and the terahertz group including Radoslaw Piesiewicz and Thomas Kleine-Ostmann for introducing me to new fields of research, and Norman Krumbholz and Christian Jördens to give me the opportunity to work together in various areas of terahertz topics. I would like to acknowledge the unlimited support and nice cooperation with Christian Jansen. His ideas, opinions, and discussions were always beneficial. My thanks are due to the other PhD students who have helped me in different ways: Mohammed Salhi, Tomasz Hasek, Kamran Ezdi, Mohammad Khalid Shakfa, Nico Vieweg, Steffen Wietzke and Maik Scheller. Thanks to the other team members for fruitful discussions and their constructive remarks. I am deeply grateful to Tobias Werth for the fruitful discussions in the beginning of my work in the microwave lab.

Furthermore, I am very thankful to all current and past IHF members, especially Ms. Christa Vogel, Ms. Kornelia Nowack and Carola Baaske for their unlimited help in administrative and bureaucratic issues. The kindness, morning greetings, helping with the library issues and willing support from Mr. Rudolf Goerke are indeed very appreciated. I would like to acknowledge the financial support of the "Deutscher Akademischer Austausch Dienst" and the International Office at Technische Universität Braunschweig during my PhD studies. Finally, my sincere thanks to my family, who offered me the support and patience along the way.

Braunschweig, 3. November 2009   Ibraheem Al-Naib

Contents

1 Introduction

Since the origin of physics, scientists were concerned with the interaction of electromagnetic waves with matter. Recently, the invention of artificial materials, which consist of periodically arranged, resonant, metallic sub-wavelength elements, enabling custom-tailored dielectric and/or magnetic responses within a certain frequency band. Today, many applications benefit from the unique electromagnetic properties that such artificial materials, also called metamaterials, offer. Especially planar metamaterials, which are easily facilitated into existing microwave circuitry, are of high practical interest, e.g. for high performance filters, antennas, and other microwave devices. Aside from the device performance, miniaturization is a key issue in the design of metamaterial resonators, as a high integration density is a mandatory prerequisite to compete in mass-markets such as wireless communications.

In fact, considerable efforts have been made towards achieving low-power miniaturized radio frequency components. However, issues related to design and fabrication of efficient, miniaturized, and easily integrable passive microwave components appear to be contradictory. Many unit cells are needed in order to achieve high efficiency components, such as filters and couplers, whose dimensions are comparable or larger than a wavelength. This wavelength dependence originates from the fact that all of the aforementioned components are composed of one or more resonant structures. Hence, high performance devices tend to be electrically large.

The main objective of this dissertation is to design and realize new resonator structures to meet the challenges of simple, high performance, and miniaturized microwave devices as well as terahertz components for filtering and sensing applications. As far as miniaturization is concerned, at least three main techniques have been proposed in the literature: lumped-element filters, high temperature super conducting (HTS) filters, and slow-wave distributed resonators. Despite the fact that filters using lumped-elements can be made very small, their insertion loss becomes unacceptable at UHF and beyond. To overcome the problem of insertion loss and probably power handling, HTS technology has been proposed. It gives an excellent solution if the cost issue will be resolved. However, superconductivity has not yet been demonstrated at room temperature. Hence, in addition to the cost and complexity associated with the HTS circuits, cooling systems and their power requirements make the current HTS technology not suitable for wireless systems. Distributed element techniques demonstrate much better performance with regard to the insertion loss and power handling capabilities compared to lumped-element filters. However, size and complexity are two major disadvantages of distributed element filters.

The dominant idea throughout this work is to propose novel and efficient miniaturized resonators for high efficiency filter and sensing applications at microwave and terahertz frequencies. The dissertation is organized as follows: chapter two will review the early steps of metamaterials and give an overview of their potential applications. Moreover, different approaches for microwave metamaterials will be presented. Finally, the basics of microwave filters, dispersion engineering, and metasurfaces will be briefly discussed. Chapter three is devoted to the description of simulation techniques and measurements procedures which were used throughout this dissertation.

In chapter four, a review of some recently proposed planar metamaterial resonator concepts will be given, illuminating their strengths and weaknesses in comparison to existing approaches. The focus will be on structures integrated on coplanar waveguides as this technology offers some distinct design advantages compared to conventional microstrip lines, e.g. the easy realization of shunts and the possibility of mounting active and passive lumped components. The first concept which falls under this category is the complementary split ring resonator (CSRR) with and without bandwidth modifying slots. These resonators provide a pronounced stopband characteristic, which can be adjusted by modifying the nearby slot lengths, offering a great design flexibility. However, the stopband response suffers from a spurious rejection band close to the main resonance, which has its origin in the differing electrical lengths of the inner and the outer resonator arm. This issue leads to the complementary u-shaped split resonator, in which the spurious resonance is suppressed by equalizing the electrical length of both resonator arms. Furthermore, the concept of achieving compact bandpass filters by combining a CSRR and a split ring resonator with strip line in series will be introduced. This approach could be very useful, e.g. in front-end filter designs. Apart from the CSRR based structures, rectangular multiple turn complementary spiral resonators will be also discussed, which enable extremely small electrical footprint filters as each turn elongates the effective resonator length, thus lowering the resonance frequency.

Although microwave filters are often physically very compact, they can contribute a lot to the delay in microwave circuits. Therefore, the study of the group delay is very important in radar and communication systems. Moreover, group delay issues are crucial in wireless communication systems. The bit rate of such systems is highly affected by the group delay of the system's components. Therefore, studying the group delay behavior for microwave filters is quite important from a system perspective. Chapter five presents a combined theoretical and experimental study on the group delay behavior of coplanar waveguide structures loaded with pairs of split-ring resonators and strip lines.

While the last two chapters discussed various metamaterial resonators, the metamaterial research has also induced advances in related fields such as frequency selective surfaces which are also known as metasurfaces or planar metamaterials. They are used in microwave and optical filters or in thin-film sensing. A variety of concepts have been introduced in the literature, but the proposed techniques still suffer from different limitations. In chapter six, rectangular asymmetric double split resonators (aDSRs) with sharp tips are presented. They offer a very high sensitivity at very small scales. While aDSRs provide a stopband behavior, many applications require just the dual response, i.e. a passband characteristic. Hence, it will be demonstrated that by applying Babinet's principle to aDSRs we can appeal to this need. Finally, asymmetric single split rectangular resonators are introduced. This resonator concept offers very high Q-factors well above 200, thus presenting an order of magnitude improvement compared to aDSRs, while even further reducing the electrical unit cell size. Finally, the conclusion are given in chapter seven.

2 Metamaterials: Fundamental Revolution and Potential Future

Materials' properties have troubled scientists since old ages [1]. From an electromagnetic outlook, researchers have had different concerns about materials' parameters and have looked at the problem from different viewpoints. While at microwave frequencies the relative dielectric permittivity is of interest, in optics the refractive index (n) is the important parameter. Usual optical materials have a positive dielectric permittivity (ϵ) and magnetic permeability (μ), and n could easily be taken as $\sqrt{\epsilon\mu}$ without any problems. Although it was realized that the refractive index would have to be a complex quantity to account for absorption and even a tensor to describe anisotropic materials, the question of the sign of the refractive index did not arise until the late sixties of the last century [2].

In 1968, Veselago studied analytically a medium that has both negative ϵ and negative μ [3]. He deduced that the medium possesses a negative refractive index. This means the negative square root $n = -\sqrt{\epsilon\mu}$, should be chosen. His result remained an academic curiosity for a long time, as neither real nor artificial materials with simultaneously negative ϵ and μ were available.

However, in the last few years, theoretical studies [4, 5] for specific engineered media whose ϵ and μ could become negative in certain frequency ranges were developed experimentally [6, 7], and this has brought Veselago's result into the limelight. These materials have been called metamaterials (MTMs). It means the materials which are not available in nature. Moreover, negative index media (NIM), double negative media (DNG), backward media, and left-handed media (LHM) have been named for the media when both ϵ and μ are negative. However, nowadays metamaterials are considered to be any engineered structures with unusual properties not readily available in nature [1]. Hence, negative permittivity, negative permeability, less-than-one refractive index, ϵ and/or μ near zero, graded index and bi-anisotropy materials are just some representative examples of metamaterials [8]. This field has become a hot topic of scientific research and debate over the past nine years. This chapter will review the early steps of the subject, and then give an overview of the potential applications.

2.1 Metamaterials - The Story so Far

2.1.1 How the Subject Started?

For ordinary materials both the relative permittivity and permeability are positive values and larger than one. These materials have been identified as right-handed media (RHM) or dou-

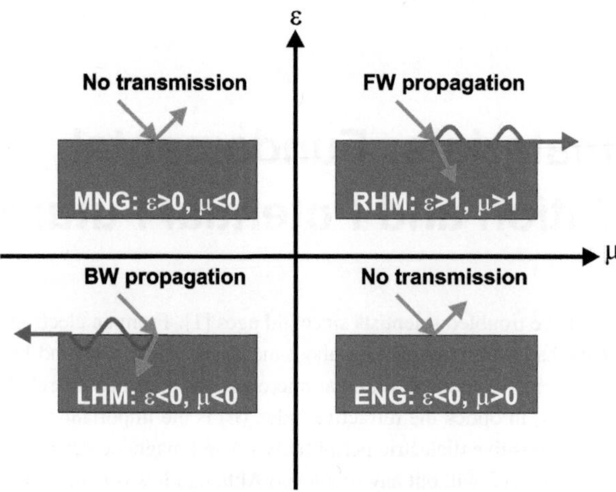

Figure 2.1: Material classification.

ble positive media (DPS). They are represented by the right-top quadrant of Fig. 2.1. In the following, some of the historical milestones for the other quadrants will be discussed.

As in any scientific field, metamaterials have required simultaneous efforts over the last century. It had a long gestation period and many contributors. The first study of general properties of wave propagation in negative index medium, represented by the left-bottom quadrant of Fig. 2.1, has usually been attributed to the work of Russian physicist V. G. Veselago. However, the subject has been studied since at least 1904 [9]. Some historical milestones regarding backward waves and metamaterials have been made over the last century. Fig. 2.2 shows the names versus years for some known researchers who contributed effectively to make that. H. Lamb in 1904 may have been the first person to propose the existence of backward waves. The phase of these waves moves in the direction opposite from that of the energy flow [10]. Lamb´s examples involved mechanical systems rather than electromagnetic waves. Apparently, the first person who discussed the backward waves in electromagnetism was Schuster in 1904 [11]. Schuster briefly annotates Lamb´s work and gives a speculative discussion of its implications for optical refraction. He cited the fact that within the absorption band of, for example, sodium vapor a backward wave will propagate. He was pessimistic about the applications of negative refraction because of the high absorption region in which the dispersion is reversed. Meanwhile, H.C. Pocklington showed in 1905 that in a specific backward-wave medium, the output wave from a suddenly activated source has group velocity which is directed away from the source, while its velocity moves toward the source [12].

After about half a century, Mandelshtam made the earliest known speculation on negative refraction [13]. He noticed that given two media, for a given incidence angle $\theta 1$ of the wave at the interface, Snell's law admits mathematically two solutions: not only the conventional solution $\theta 2$ but also the "unusual" solution π-$\theta 2$. Although we note that he made no reference

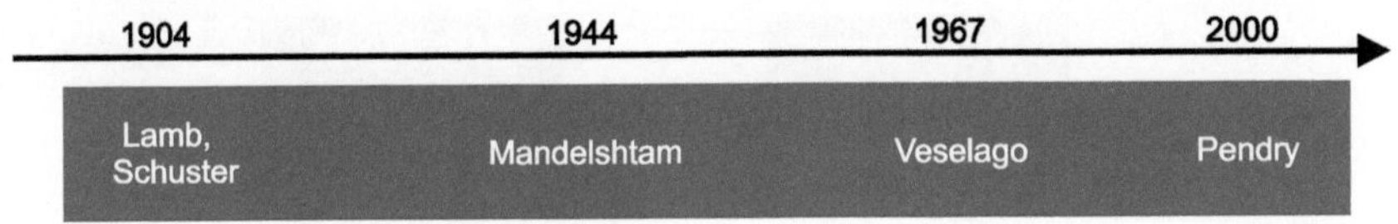

Figure 2.2: Main historical milestones of metamaterials.

to negative refraction, it is obvious that his argument refers to the same phenomenon. In 1945, with reference to Lamb, Mandelshtam presented physical examples of structures supporting waves with "negative group velocity". Such structures exhibit periodically varying effective permittivity [14]. From the above reference with Mandelshtam, it is concluded that Veselago was not the first to postulate the existence of LH media in [3]. However, it is an established fact that he was the first to conduct a systematic study of these media and to predict their most fundamental properties. Moreover, Pendry's contributions have awakened scientists to the aforementioned physical phenomenon [4, 5, 15].

2.1.2 Wave Propagation in Left-Handed Media

Let us start with the wave equation to show wave propagation in left-handed media [3]:

$$\left(\nabla^2 - \frac{n^2}{c^2}\frac{\partial^2}{\partial t^2}\right)\psi = 0 \tag{2.1}$$

where n is the refractive index, c is the velocity of light in vacuum, and $n^2/c^2 = \epsilon\mu$. It is anticipated then that lossless left-handed media with $n = -1$ must be transparent [16]. Considering the above equation, we can interestingly conclude that solutions to equation (2.1) will remain unchanged after a simultaneous change of the signs of ϵ and μ. However, when Maxwell's first-order differential equations are explicitly considered,

$$\nabla \times \mathbf{E} = -jw\mu\mathbf{H} \tag{2.2}$$

$$\nabla \times \mathbf{H} = jw\epsilon\mathbf{E} \tag{2.3}$$

where $\mathbf{E}$ is the electric field and $\mathbf{H}$ is the magnetic field, it becomes obvious that these solutions are quite different. For plane-wave fields, the above equations reduced to:

$$\mathbf{k} \times \mathbf{E} = w\mu\mathbf{H} \tag{2.4}$$

$$\mathbf{k} \times \mathbf{H} = -w\epsilon\mathbf{E} \tag{2.5}$$

where $\mathbf{k}$ is the wave vector. Therefore, for positive ϵ and μ, $\mathbf{E}$, $\mathbf{H}$, and $\mathbf{k}$ form a right-handed system of vectors as shown in Fig. 2.3(a). However, if $\epsilon < 0$ and $\mu < 0$, then equations (2.4) and (2.5) can be rewritten as:

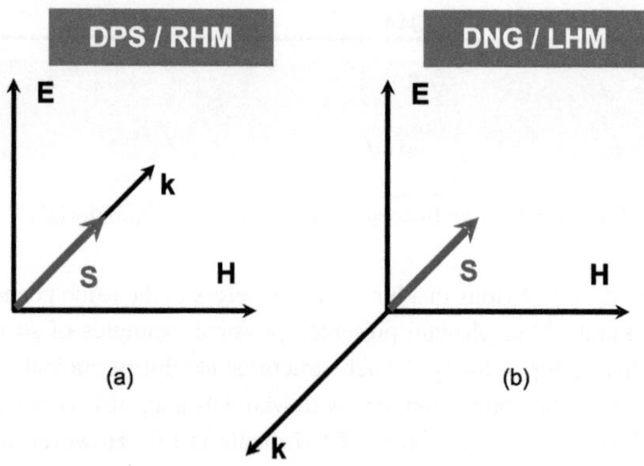

Figure 2.3: Representation of the three vectors **E**, **H** and **k** for a right-handed medium (RHM) (a) and left-handed medium (LHM) (b).

$$\mathbf{k} \times \mathbf{E} = -w|\mu|\mathbf{H} \tag{2.6}$$

$$\mathbf{k} \times \mathbf{H} = w|\epsilon|\mathbf{E} \tag{2.7}$$

showing that **E**, **H**, and **k** now form a left-handed triplet, as illustrated in Fig. 2.3(b). In fact, this result is the original reason for the denomination of negative ϵ and μ, media as "left-handed" media [3].

The main physical implication of the aforementioned analysis is backward-wave propagation. For this reason, the term backward media has been also proposed for media with negative ϵ and μ [17]. In fact, the direction of the time-averaged flux of energy is determined by the real part of the Poynting vector,

$$\mathbf{S} = \frac{1}{2}\mathbf{E} \times \mathbf{H}^* \tag{2.8}$$

where * denotes the complex conjugate, which is not influenced by a simultaneous change of sign of ϵ and μ. Thus, **E**, **H**, and **S** still form a right-handed triplet in a left-handed medium. Hence, energy and wavefronts travel in opposite directions in such media. Backward-wave propagation in homogeneous isotropic media seems to be a unique property of left-handed media. The phenomena of forward- and backward-wave propagation are demonstrated in Fig. 2.4(a) and Fig. 2.4(b) for a slab of media with refractive index of 1 and -1, respectively. There is free space before and after the media. The backward wave is evident from the electric field distribution inside the media in Fig. 2.4b. So, the wave vector is opposite of the pointing vector.

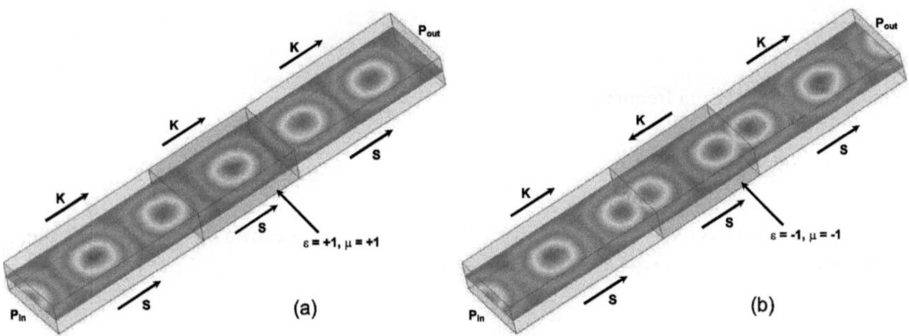

Figure 2.4: Forward- (a) and backward- (b) wave demonstration using a slab of material with ϵ and μ equal 1 and -1, respectively.

2.1.3 Negative Permittivity

It is not secret that higher permittivity low loss materials were needed during the Second World War for radar technology. Therefore, there were great efforts made to develop artificial dielectrics. One of the structures studied was an array of thin wires, which were shown to have an effective plasma frequency [18]. Later, Rotman was motivated to simulate plasmas in order to have more insight into problems such as the effect of rocket exhaust upon the radiation of re-entry vehicle antennas and his paper [19] was a result of that investigation. About quarter of a century later, Pendry et al. obtained similar conclusions [4]. The idea is based on having an array of these wires as shown in Fig. 2.5(a) with an applied electric field along the axes of the wires.

It has been shown that the behavior of an array of thin metallic wires can be explained by the plasma resonance in a metal [4, 20]. The frequency response of a metal to incident electromagnetic radiation is due to the plasma resonance of the electron gas. The following expression describes this behavior in an ideal form:

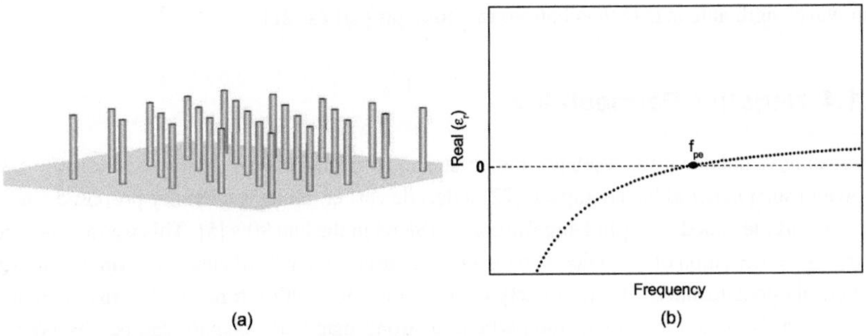

Figure 2.5: (a) Wire medium with an applied electric field along the axes of the wires (b) real part of ϵ_r versus frequency.

$$\epsilon_{metal} = 1 - \frac{w_{pe}^2}{w^2} \tag{2.9}$$

where w_{pe} is the plasma frequency, which is given by:

$$w_{pe}^2 = \frac{ne^2}{\epsilon_0 m_e} \tag{2.10}$$

where, n: electron density of the gas, m_e: electron mass, e: electron charge. An analytical approximation of the thin wire medium shows that:

$$n_{eff} = n\frac{\pi r^2}{a^2} \tag{2.11}$$

$$m_{eff} = \frac{\mu_o \pi r^2 e^2 n}{2\pi} ln\left(\frac{a}{r}\right) \tag{2.12}$$

where n_{eff} and m_{eff} are the average electron density and the effective electron mass, respectively. r and a are the wire radius and the lattice period, respectively. Simple substitutions give the plasma frequency:

$$w_{pe}^2 = \frac{2\pi c_0^2}{a^2 ln\left(\frac{a}{r}\right)} \tag{2.13}$$

By applying equation 2.13, a plot of epsilon for a thin wire medium can be calculated, as is shown in Fig. 2.5(b). The graph explains how epsilon goes from negative values in the lower frequency range (and hence, only evanescent modes are allowed to propagate), up to positive values in the higher frequency range, through the plasma frequency f_{pe}. It is also interesting to note that the plasma frequency can be lowered by increasing the lattice constant (a) of the medium. Fig. 2.6 shows a sweep for four different ratios for a/r. As the ratio increases, the plasma frequency decreases. This is an interesting property of this medium, because the operating frequency range of the final metamaterial configuration is limited by the size of this metallic array. Researchers have been attracted by this kind of thin wire arrays. They have envisaged these types of structures with plasmonic response for the realization of sub-wavelength antennas with enhanced radiation properties [21].

2.1.4 Negative Permeability

Interestingly, unknown to Veselago, the existence of negative permeability had already been shown in such material by Thompson [22] a decade earlier. However, Pendry proposed a novel type of particle called the Split Ring Resonator (SRR) in the late 90´s [5]. This was a major step in the implementation of an LHM. The resonator consists of a pair of concentric rings, with slits etched in opposite sides. By adequately exciting the SRR with a time varying magnetic field oriented in the axial direction of the particle, a strong magnetic behavior can be observed. A highly resonant response with frequency can be observed by studying the equivalent magnetic permeability value. Moreover, a frequency range will exist in which μ will exhibit a very high

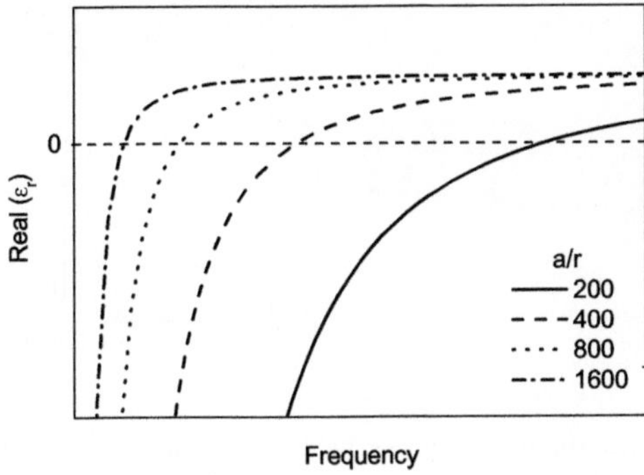

Figure 2.6: Analytical calculation of epsilon for a thin wire medium when a/r ratio is varied.

negative value (close to the quasi-static frequency) up to where it reaches a value higher than 0 (magnetic plasma frequency).

Pendry started his study by taken two concentric cylinders with a slit in each one, in opposition one to the other. A schematic of the resulting structure is shown in Fig. 2.7. In this case, currents are forbidden to move in any of the cylinders. A very high capacitive value is a consequence of that, which enables displacement current to flow. The value of the effective magnetic permeability is given by:

$$\mu_{eff} = 1 - \frac{F}{1 + i\frac{2\sigma}{wr\mu_0} - \frac{3}{\pi^2\mu_0 w^2 C r^3}} \tag{2.14}$$

where F is the fractional volume, given by $\pi r^2/a^2$ and C is the capacitance per unit area between both metallic sheets of the cylinders, given by $C = \epsilon_0/d = 1/dc_0^2\mu_0$. Fig. 2.8 presents the SRRs medium and the analytical calculation of Eq. (2.14) for the effective magnetic permeability of a medium composed by SRR cylinders. It shows the μ_{eff} versus frequency.

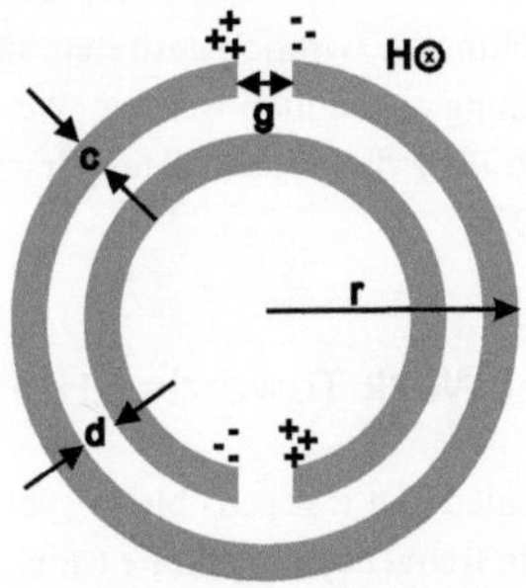

Figure 2.7: Top-view of split ring resonator cylinder structure.

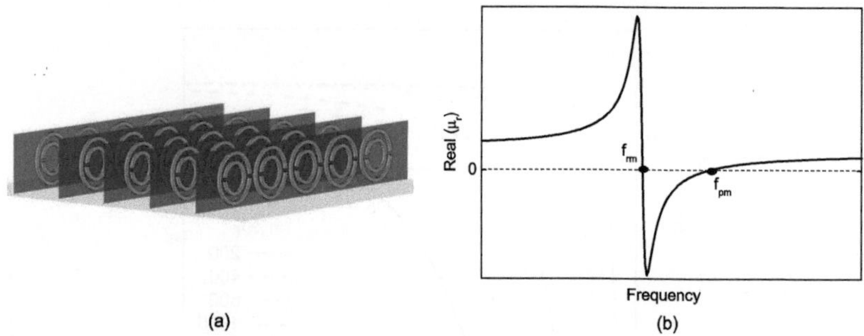

Figure 2.8: (a) SRRs medium with magnetic field along their axis (b) analytical calculation of the effective magnetic permeability value for SRR cylinders.

The effective magnetic permeability presents a highly resonating peak at the so-called quasi-static resonance frequency, f_{rm}. At that point, the magnetic permeability becomes highly negative, being less negative as frequency is increased. The point at which the magnetic permeability equals zero is known as the magnetic plasma frequency f_{pm}. Therefore, the range in which the magnetic permeability is negative is:

$$f_{rm} < f_{\mu<0} < f_{mp} \tag{2.15}$$

It is worth mentioning that there was a glance at the history of SRR even in Hertz's time. He wrote in one of his books "Untersuchungen über die Ausbreitung der elektrischen Kraft" in German language the following: Die elektrischen Oscillationen geöffneter Inductionsapparate haben eine Schwingungsdauer, welche nach Zehntausendtheilen der Sekunde gemessen werden kann. Etwa hundertmal schneller erfolgen die Schwingungen oscillirender Flaschenladungen, welche Feddersen beobachtete. **Schnellere Schwingungen noch als diese lässt die Theorie als möglich voraussehen in gutleitenden ungeschlossenen Drähten, deren Enden nicht durch grosse Capacitäten belastet sind, ohne dass freilich die Theorie zu entscheiden vermöchte, ob solche Schwingungen je in bemerkbarer Stärke thatsächlich erregt werden können.** Gewisse Erscheinungen legten mir die Vermuthung nahe, dass Schwingungen der letztgenannten Art unter bestimmten Verhältnissen wirklich auftreten, und zwar in solcher Stärke, dass ihre Fernwirkungen der Beobachtung zugänglich werden. Weitere Versuche bestätigten meine Vermuthung, und es soll deshalb über die beobachteten Erscheinungen und die angestellten Versuche hier berichtet werden [23].

2.1.5 First Experimental Work Towards LHM

The preceding sections have revealed that it is possible to synthesize artificial particles, which exhibit negative values in a certain frequency range for ϵ (thin conductor wire medium) and for μ (SRR particles). The logical next step is to combine both structures to build a LHM in a single configuration. Therefore, the first experimental observation of LHM was proposed by Smith et.

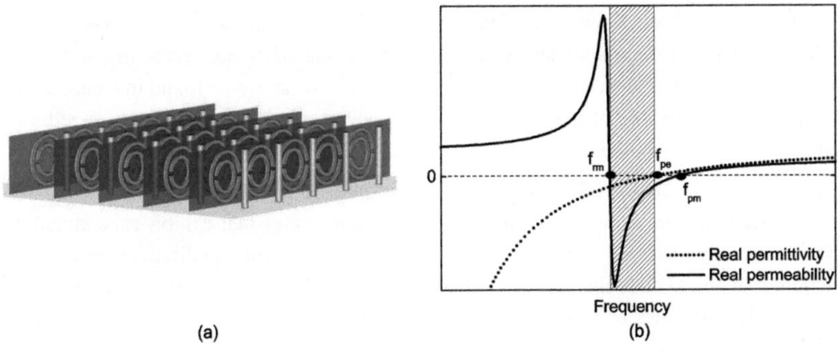

Figure 2.9: (a) Schematic of the LHM, resulting from the combination of an array of SRR particles with an array of thin wires. (b) Real part of effective permittivity (dotted) and permeability (solid) versus frequency.

al [6, 7]. Fig. 2.9(a) depicts a schematic of the final structure, which consists of an array of SRRs which are placed between an array of metallic posts.

The experimental setup comprised an array of SRRs with an array of thin wire structure placed in a scattering chamber. The results were in good agreement with the analytical expectations. When the SRR medium was measured alone, a stopband appeared in the frequency response. Only an evanescent wave can propagate in this case. However, if the wire medium was added, a passband could be measured in the same frequency range where previously rejection was detected. This result is interpreted to the combination of simultaneously $\epsilon < 0$ and $\mu < 0$ allowing propagation since the Poynting vector once again exists and is real. Besides being the first experimental validation of an LHM, the work pointed out two important facts:

1. The frequency response of the structure is modified by the field incidence in the composite medium. This behavior is obtained only if the **H**-field is parallel to SRR axis.

2. The fractional area F (relation between the SRR radius and the cell lattice) is a key parameter in the enhancement of the permeability.

However, this structure has many drawbacks such as the complexity and matching to free space. Within an ongoing interest, some other approaches have been proposed for the implementation of LHM, such as chiral medium based [24] and composite structures composed of thin metallic plates or wires embedded in a ferromagnetic substrate [25].

2.2 Definition of Metamaterials

In science, words are thought to mean roughly the same thing to different people [1]. However, it is not absolutely correct as far as metamaterials definition is concerned because, as was mentioned earlier, the interest in this subject has greatly increased recently. Sihvola has discussed this matter in detail [1]. So, the aim here is simply to give brief summary.

The Metamorphose Network of Excellence [26] by the European Union has formalized MTMs as: Artificial electromagnetic (multi-) functional materials engineered to satisfy the prescribed requirements. Superior properties as compared to what can be found in nature are often underlying in the spelling of metamaterial. These new properties emerge due to specific interactions with electromagnetic fields or due to external electrical control.

In the USA, the DARPA Technology Thrust program on metamaterials [27] defined MTMs as: MetaMaterials are a new class of ordered nanocomposites that exhibit exceptional properties not readily observed in nature. These properties arise from qualitatively new response functions that are: (1) not observed in the constituent materials and (2) result from the inclusion of artificially fabricated, extrinsic, low dimensional inhomogeneities.

Also characterizations exist for metamaterials that use fewer words and attempt to pinpoint the most essential property of these materials, such as the one in Ref. [28]: Materials made out of carefully fashioned microscopic structures can have electromagnetic properties unlike any naturally occurring substance. And web page of Professor David R. Smith [29]: Artificially structured metamaterials can extend the electromagnetic properties of conventional materials.

Two fundamental properties can be noticed from the listed definitions. Metamaterials should exhibit properties:

- not observed in the constituent materials,

- not observed in nature.

These two properties point to the heart of the various ways metamaterials are understood. Moreover, many researchers are of the opinion that the great umbrella of metamaterials also covers electromagnetic crystals, which are known by many other names: photonic crystals, electromagnetic band gap structures (EBG), and photonic bandgap structures (PBG) [1].

2.3 Hot Applications of Metamaterials

The metamaterials approach has been also successfully applied to design a variety of new structures with unique electromagnetic properties.

1. Superlens: Pendry proposed that a slab of LHM can be used as a lens which is free from all aberrations observed in a lens made with positive refractive index as shown in Fig. 2.10 [15]. However, it was shown that very small deviations of the material parameters from the ideal conditions could lead to the excitation of resonances that cause deterioration of the performance of the lens. Nevertheless, scientists have been working to overcome different difficulties to improve the resolution [30]. It is worth mentioning in this regard that the oldest lens ever found, by Sir John Layard in 1850, was at the palace of Nimrud in what is now Iraq. Fig. 2.11 presents the Nimrud lens. It had been found in deposits dated around 600 *BC* and although its provenance was not in question, doubts were raised about its function. Whilst it clearly works as a lens, it was thought to have been used as a decoration in a piece of jewellery [31].

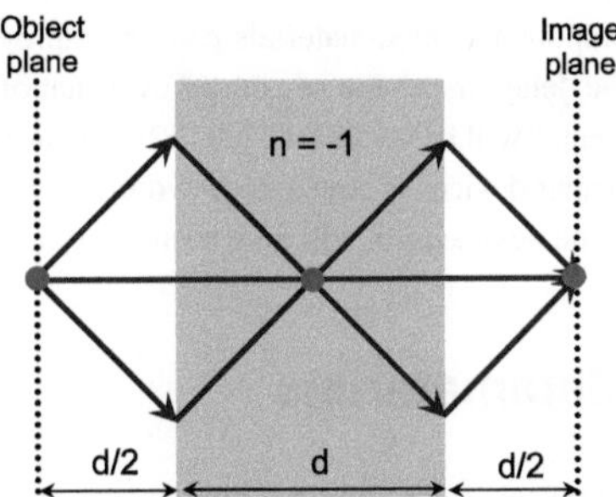

Figure 2.10: Schematic of the focusing phenomenon in LHM.

2. Cloaking: Another natural application for metamaterials is the development of gradient index media [32] because the value of the permittivity and permeability can be engineered at any point within the structure by adjusting the scattering properties of each unit cell [33, 34]. By implementing complex gradients independently in the permittivity and permeability tensor components, it has been shown that an entirely new class of materials can be realized by the process of transformation optics [35, 36]. A recent example utilized metamaterials to form an "invisibility cloak" that was demonstrated to render an object invisible to a narrow band of microwave frequencies [37].

3. Scattering reduction: metamaterials can be used for the reduction of electromagnetic wave scattering [38, 39]. Recently, a theoretical analysis based on Mie scattering was presented in [38, 39] which indicated that when a metal is coated with metamaterials the scattering coefficient would be very small.

4. Optical receivers using EBG based analogous LHM: The spreading of the illumination area in the focal plane of a lens can be suppressed by using negative refraction behavior of photonic crystals. In [40], a field of view expander (FOVE) technique based on this approach for optical receivers was proposed. In optical detectors, the light acceptance area is limited so that a reduction of the focal spot's illumination area at the reception of an aberrated beam will take place. The propagation angles of the received beam in front of optical devices need to be taken into account. The proposed method in [40] can significantly reduce the focal spot size without expanding the propagation angles.

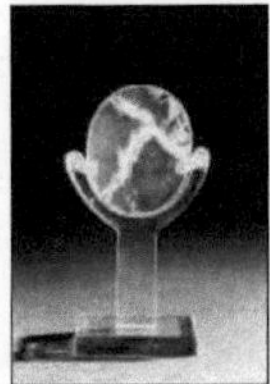

Figure 2.11: The Nimrud lens.

5. Novel microwave components: metamaterials can be employed as sub-wavelength resonators and zero phase delay lines. The advantage over that of RHM materials is that the dimension of the resonator will be very small [41, 42]. As far as application of metamaterials for microwave planar devices is concerned, two main concepts have been developed over the last decade. The next section will give a summary of these concepts.

2.4 Microwave Metamaterials

The aforementioned early steps to construct metamaterials were significant steps, but practical applications are hard to find for such elaborate three dimensional structures. In order to bring metamaterial technology to microwave components, compatibility with planar circuit technology is mandatory. Two approaches have been introduced to meet this challenge: The first one employs a transmission line and the second one is a metamaterial resonator loaded coplanar waveguide planar technology.

2.4.1 Transmission Line Approach

In this approach, a transmission line (TL) is loaded with reactive elements, such as capacitors and inductors, so that the effective series impedance becomes capacitive while the effective shunt impedance remains inductive. The main advantage of this concept is its compatibility with conventional planar circuits. Fig. 2.12(a) illustrates the transmission line model of a TL-based metamaterials and Fig. 2.12(b) demonstrates one of the proposed implementations [43].

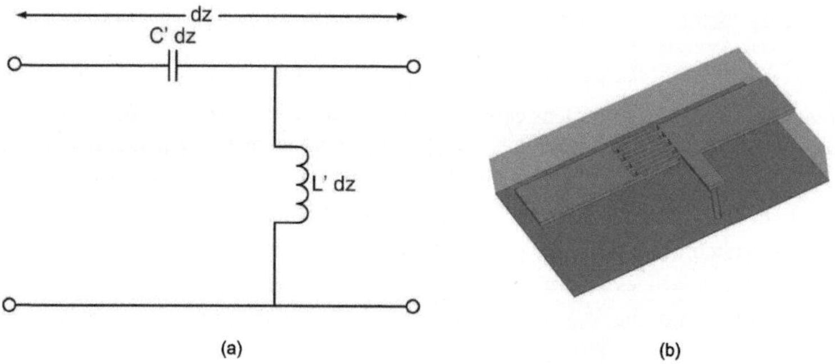

Figure 2.12: Transmission line model (a) for the TL-based metamaterial configuration (b).

Therefore, if a transmission line is loaded with reactive elements in such a way that the effective series impedance value is capacitive and the effective shunt value is inductive, then a consequence response of an LHM can be obtained. The main advantage of this concept is enabling its integration in conventional planar circuits, which is of widespread. Several practical implementations of LHM based on a TL approach have been proposed and experimentally validated.

2.4.2 CPW-SRRs Approach

The coplanar waveguide offers a variety of advantages over the conventional microstrip line. The most important benefit is that the CPW facilitates easy shunt as well as series surface mounting of active and passive devices. Another advantage is that the effective CPW characteristics can be easily modified by introducing slotlines which are coupled to the CPW. Then the CPW appears as if it was a CPW with circuit elements connected to it in series or in parallel. Advantages of the slotline configuration are low cost, light weight, and small size. Consequently, it has led to several applications in microwave integrated circuits (MICs). These, as well as several other advantages make the CPW ideally suited for MICs and many other applications [44, 45, 46]. Fig. 2.13 visualizes the CPW with the transverse electric and magnetic fields.

It has been shown in the preceding sections that SRR particles can be employed to synthesize LHM in free space, if appropriate excitation of the particles is utilized. Hence, it is possible to do that with planar components if the H-field is normal to the plane that contains the SRRs. By maintaining this condition, the induced currents will lead to the desired negative value of the magnetic permeability. Fortunately, H-field incidence lines in CPW, which are perpendicular on its interface (see Fig. 2.13b), give a hint on the possibility of exciting SRR particles adequately.

To do so, the SRR particles must be somehow included in the structure of the waveguide. To maintain the main advantage of having a single metal layer in the CPW, the first approach was by introducing of the particles on the slots of the CPW as depicted in Fig. 2.14(a) which has been proposed in Ref. [47] for which we only show simulations. It consists of one metal layer only. Note that a device with only a single metal layer is easier to fabricate than a structure based on two metalized layers. The gaps of the CPW are broadened to provide enough space to hold the SRRs. The SRRs are excited magnetically because the magnetic field is concentrated in the gaps. Fig. 2.14(b) shows the transmission parameters numerically obtained. The transmission depicts a bandstop behavior due to the magnetic resonance frequency. Observing the return loss, it is noticed that the performance is incredibly poor. This leads to unacceptable return loss because the structure is highly mismatched.

This allows integration of the SRRs within the host CPW. However, CPW line impedance

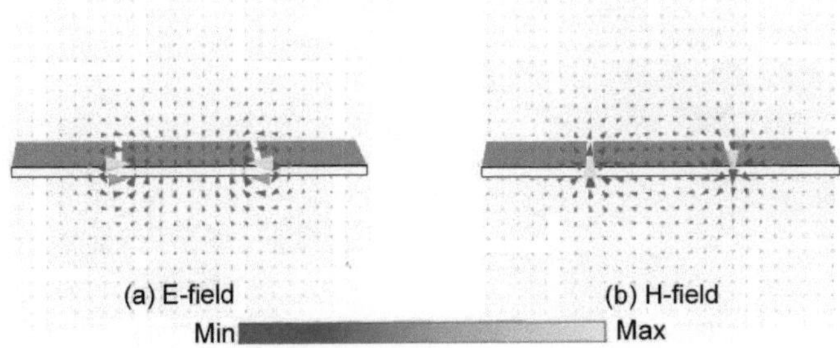

Figure 2.13: The transverse electric- (a) and magnetic- (b) field in CPW.

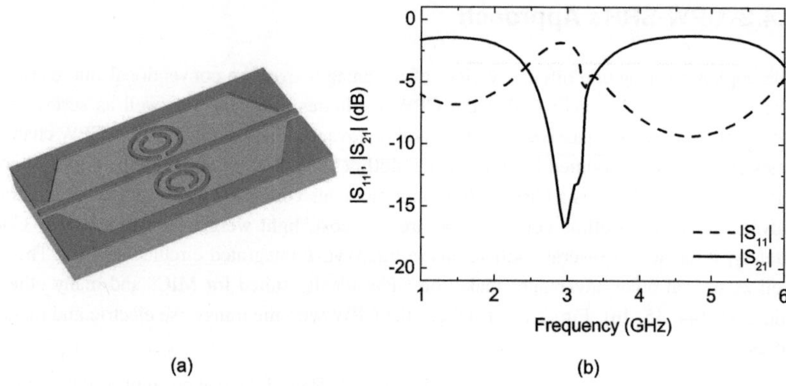

(a)   (b)

Figure 2.14: Single metallized CPW with SRRs inside the slots (a) and its S-parameters (b).

is the ratio between the central conductor width and the air gap separation. Therefore, having the SRRs inside the slots of CPW will put high restriction to choose the line impedance. Hence, only single metal CPW metamaterials structures are considered impractical. To overcome this difficulty, another approach has been proposed. The SRRs are adequately placed in the bottom side of the dielectric layer as shown in Fig. 2.15(a) [48]. In this case, CPW can be tailored to have almost the same impedance as if no SRRs were present. The insertion loss (See Fig. 2.15(b)) revealed very sharp response with four unit cells of SRRs. However, the main advantage of having a single metal layer is no longer maintained. Besides, Falcone et al. proposed to use complementary split-ring resonators (CSRR) in microstrip lines [49]. According to the concept of duality (Babinet principle) this resonator leads to the same transmission characteristics. In [49] the CSRR has been etched into the ground plane, underneath the upper microstrip transmission line.

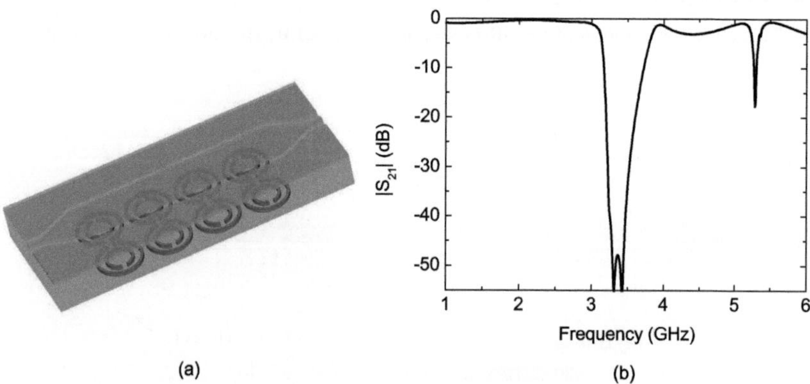

(a)   (b)

Figure 2.15: CPW with backside loaded SRRs (a) with its transmission characteristic (b).

2.5 Basics of Microwave Filters, Dispersion Engineering, and Metasurfaces

In the following, we will start with the fundamental definitions and theorems which are used throughout this dissertation. First of all, microwave filters which have been widely used in the today's communication systems are presented. After that, the dispersion analysis will be shortly introduced. Finally, the basics of conventional metasurfaces will be briefly explained.

2.5.1 Microwave Filters

Microwave filters act as an essential block in various wireless systems that operate at frequency ranges above 300 MHz. They are utilized to control different frequencies, i.e. transmitting the required signals in specific passband regions while attenuating all the undesired signals in the remaining bandstop regions. Emerging applications such as wireless communications continue to challenge RF filters with ever more stringent requirements of better performance, smaller size, lighter weight, and lower cost. Basically, filters can be classified into four different types: lowpass, highpass, bandpass, and bandstop filter as shown in Fig. 2.16. The first two types realize transmission in the frequency band below or above a cutoff frequency while the third and fourth types transmit or attenuate signals within a certain frequency band that is terminated by the lower and upper cutoff frequencies.

Filters could be designed as lumped element or distributed element circuits depending on the requirements and specifications. In general, the lumped-element filter design has good performance at low frequencies. However, two problems arise at microwave frequencies. First, lumped elements such as inductors and capacitors should be approximated with distributed components because they are available only for a limited range of values. In addition, at microwave

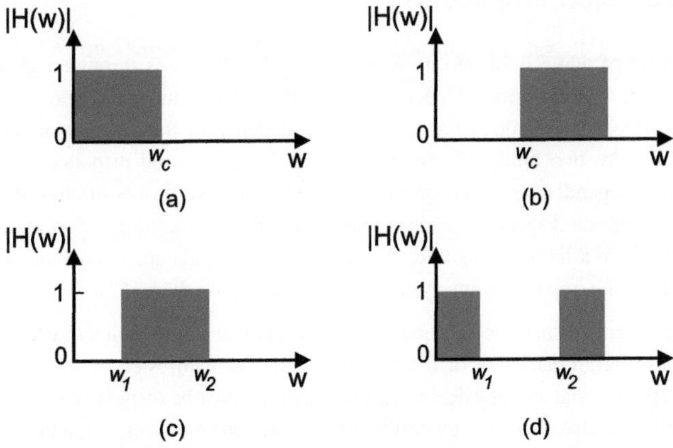

Figure 2.16: Four types of filters: (a) Low-pass filter (b) high-pass filter(c) bandpass filter (d) bandstop filter.

frequencies the distances between filter components is not negligible. In contrast, the distributed element filters may be realized in various transmission line structures, such as waveguide and microstrip. Therefore, they could work at higher frequencies. Moreover, recent advance in novel materials and fabrication technologies including MMIC and MEMS have stimulated the rapid development of new microstrip filters. Meanwhile, advances in computer-aided design (CAD) tools such as full-wave electromagnetic (EM) simulators have revolutionized filter design. Hence, many microstrip filters with advanced filtering characteristics have been proposed. However, the presence of spurious bands is an important limitation of microwave filters implemented by means of distributed elements [50]. These undesired frequency bands can seriously degrade filter performance and may be critical in certain applications that require huge rejection bandwidths. Unfortunately, for most filter implementations the first spurious band is relatively close to the frequency region of interest.

Several compact and high performance components have been suggested by employing the generic structure known as electromagnetic bandgap (EBG) or alternatively called photonic band gap (PBG) structures. These structures have been considered to obtain the required function along with circuit size reduction. However, it is not easy to use PBG structures for the design of the microwave or millimeter wave components due to the difficulties of the modeling. There are many design parameters, such as the lattice shape and lattice spacing. They have an effect on the bandgap property. Another difficulty in using the PBG circuit is caused by the radiation from the periodic etched defects.

Hence, improved performance, very compact size, and novel functionalities are highly desired. Metamaterials concepts offer all that due to the unique and controllable electromagnetic properties of its structures. Chapter Four reviews the novel structures that were developed in this dissertation.

2.5.2 Dispersion Engineering

The splitting-up of white light into its components when it passes through a glass prism is a familiar example of dispersion. Hence, it causes the separation of a wave into components with different frequencies, due to a frequency dependence of the velocity of the wave [51]. Basically, there are two kinds of dispersion. First is the material dispersion, which caused by a frequency dependent response of a material when a wave passes through it. The second kind is the waveguide dispersion, which arises when the transverse mode solutions for waves are confined within a finite waveguide. As far as metamaterials are concerned, studying their dispersion is of high impact for any system that might employ them [52].

Today, the term "dispersion engineering" reflects our desire to synthesize and control dispersive effects, and in particular their associated signs, as manifested by the phase and group delay [53]. The fact that any physical realizable medium must be dispersive is the consequence of the causality principle [54]. Let us start with the basic definitions regarding the dispersion and how things are related. For a one-dimensional dispersive problem, the behavior of a medium can also be described by the dependence of the propagation vector on frequency according to Ref. [55]:

$$k(w) = k(w_0) + \frac{\partial k}{\partial w}\Big|_{w_0}(w - w_0) + \frac{1}{2}\frac{\partial k^2}{\partial w^2}\Big|_{w_0}(w - w_0)^2 \cdots$$
$$= v_p^{-1}w_0 + v_g^{-1}(w - w_0) + \frac{1}{2}GVD(w - w_0)^2 + \cdots \tag{2.16}$$

In equation (2.16) the coefficients of expansion v_p, and v_g are the phase and group ve-locities, and GVD is the group velocity dispersion. With the photonic dispersion relation $k(w) = wn_p(w)/c$, the dispersive effects of equation (2.16) can be also described in terms of the index of refraction, n_p [53]. Hence, the second coefficient of expansion is the group delay, which is related to the refractive index by

$$n_g = n_p + w\frac{dn_p(w)}{dw} \tag{2.17}$$

The relations between the phase index and phase velocity, group index and group velocity are then as follows:

$$v_p = \frac{c}{n_p(w)} \tag{2.18}$$

$$v_g = \frac{c}{n_g(w)} \tag{2.19}$$

However, it is easier to deal with the transfer function of any system which is given by:

$$T(w) = |T(w)|exp[j\phi(w)] \tag{2.20}$$

Assuming the transmission function magnitude is constant over narrowband excitation, a Taylor series can be used to expand the phase of the transfer function as follows:

$$\phi(w) = \phi(w_0) + \frac{\partial \phi}{\partial w}\Big|_{w_0}(w - w_0) + \frac{1}{2}\frac{\partial^2 \phi}{\partial w^2}\Big|_{w_0}(w - w_0)^2 \cdots$$
$$= -\tau_p w_0 - \tau_g(w - w_0) - \frac{1}{2}GDD(w - w_0)^2 + \cdots \tag{2.21}$$

where the phase delay (τ_p), group delay (τ_g), and group delay dispersion (GDD) are given by:

$$\tau_p = -\frac{\phi}{w}\Big|_{w_0} \tag{2.22}$$

$$\tau_g = -\frac{\partial \phi}{\partial w}\Big|_{w_0} \tag{2.23}$$

$$GDD = -\frac{\partial^2 \phi}{\partial w}^2\Big|_{w_0} \tag{2.24}$$

The relations between the phase and group delays are feasible to spatial systems equations (2.23, 2.24). Then, the phase and group velocity is as follows:

$$v_p = \frac{c}{n_p(w)} = \frac{L}{\tau_p} \tag{2.25}$$

$$v_g = \frac{c}{n_g(w)} = \frac{L}{\tau_g} \tag{2.26}$$

Finally, two more points are worth mentioning. First, from equation (2.26) it is clear that for a physical system of length L the sign of the group velocity and group delay are the same. Second, when discussing a spatially extended system, the fundamental requirements of causality, also referred to as "primitive causality," must be augmented with relativistic causality, also referred to as "macroscopic" or "Einstein causality" [56]. Now, the notion of "Dispersion Engineering" declares our desire to synthesize and control various dispersive effects, and in particular their associated signs, as manifested by the phase delay, group delay, group delay dispersion [53].

The physical meaning of the phase and group delays are the delays encountered by sinusoidal time harmonics and the wave envelope as they propagate through the media, respectively. Under conventional propagation conditions, both the phase and group velocities are positive, implying the fact that both the sinusoidal time harmonics and the pulse envelope move away from the source. In contrast, in the case of negative phase but positive group velocity the sinusoidal time harmonics move toward the source while the wave packet envelope moves away from the source. This phenomenon is referred to as backward-wave propagation [57] as aforementioned in this chapter and is the signature of the LHM studies so far [58, 59, 60, 61]. More interestingly, it is also possible to observe positive phase but negative group velocities [62, 63, 64, 65]. Under these circumstances, the observer will note that the sinusoidal time harmonics move away from the source but the wave packet envelope moves toward the source. Chapter Five will present a combined theoretical and experimental study on the group delay behavior of CPW-LHM which can be taken as a pattern for any other structure of interest.

2.5.3 Metasurfaces

Frequency selective surfaces (FSSs) have been the subject of great interest for their widespread applications as spatial microwave and optical filters [66, 67]. Recently, they have also been proposed for use in indoor environments for improving the spectrum efficiency wireless communication systems [68, 69]. Conventional FSSs consist of periodically arranged, arbitrarily shaped metallic patches with a typical inter-element spacing and unit cell dimension in the order of half the guided resonance wavelength λ_g. Fig. 2.17 depicts a 3x3 FSS with square elements. The dashed area denotes the unit cell. For some applications, e.g. low-frequency antenna radomes or frequency selective electromagnetic interference shielding, FSS unit cells of much smaller electrical size are required [70, 71]. Such structures can be realized by employing planar metamaterial resonators and are also known as metasurfaces.

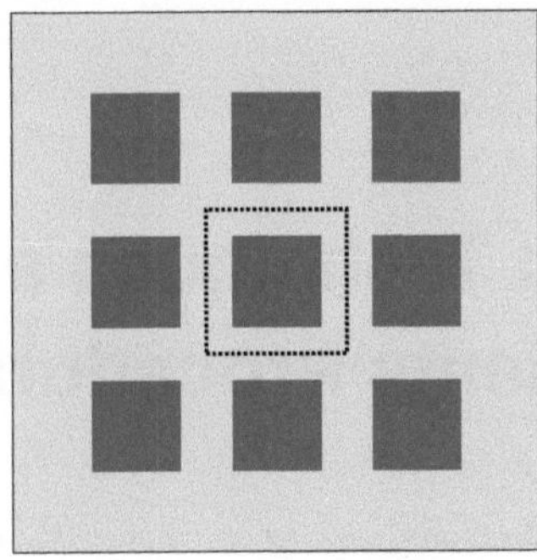

Figure 2.17: Top-view of conventional frequency selective surface.

However, FSSs suffer from many drawbacks. First of all: in such structures, the size of the resonant elements and the inter-element spacing are generally comparable to half a wavelength at the desired frequency of operation. For some applications, such as low-frequency antenna radomes or frequency selective electromagnetic interference shielding, FSSs of relatively small electrical dimensions are highly desirable.

Furthermore, FSSs designed using traditional techniques and composed of resonant elements demonstrate a bandstop or bandpass response. In situations where a high out of band rejection or sharp transmission response is required, multiple FSS panels are cascaded with quarter-wavelength spacing between each panel. Therefore, FSSs with higher-order bandpass responses are usually thick and bulky. This reduces their attractiveness for applications where conformal FSSs are required. Moreover, the quality (Q-) factor for such structures is rather low. This is a key factor where sharp response is required. Sharper resonance compared with the conventional one has been reported recently by crossing the symmetry of the double split resonator (DSR). However, the Q-factor is still not that high (about 20) and the size of the resonator is about $\lambda_g/4$ [72]. Chapter Six is devoted to review the solutions which were developed in this dissertation to overcome these potential limitations.

3 Simulation Techniques and Measurements Procedures

Necessity is the mother of innovation. However, any innovation or new idea needs to be validated. The validation requires some tools and experiments to prove the functionality of the idea. We are concerned in this work with electromagnetic and radio frequency devices. Therefore, this chapter is devoted to point out the computational electromagnetic techniques (CETs), which are used throughout this dissertation. Then, the fabrication of the structures is explained briefly. After that, the measurements procedures are demonstrated for both, planar and waveguide configurations. The chapter closes with a short summary. Fig. 3.1 demonstrates the flow chart of the above mentioned steps in order to validate any new ideas.

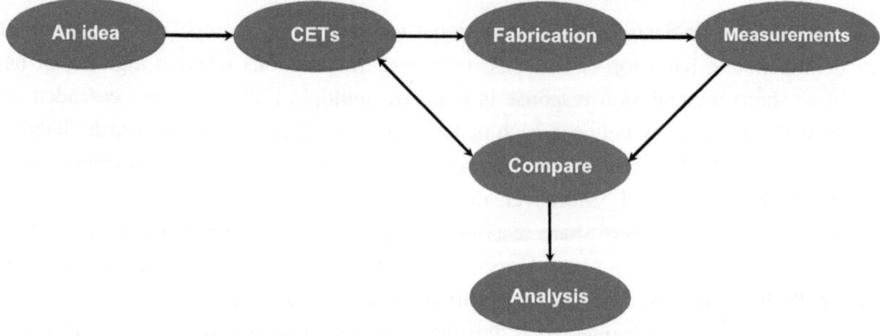

Figure 3.1: Flow chart of the required numerical and experimental steps.

3.1 Computational Electromagnetics Overview

Various mathematical concepts and tools were proposed and applied to solve Maxwell's equations for design and analysis of radio frequency components and devices. The computational electromagnetic techniques not only allow us to evaluate a device before fabricating the structure, but also provide an understanding of the performance of the device. From the point of view of research, this is a very important feature and in some cases is more valuable than the proper operation of the particular device itself. The number of techniques, algorithms and their variations engaged with computational electromagnetics has become almost uncountable.

Fortunately, all the electromagnetic computational techniques basically deal with solving Maxwell's equations at different domains. Given that the Maxwell equations can be written

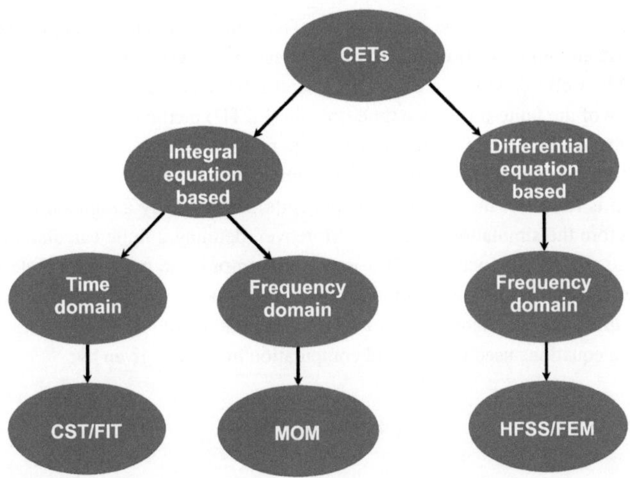

Figure 3.2: Computational electromagnetics techniques categories.

either in integral or differential forms, different computational techniques can be deployed. Fig. 3.2. shows the categories in CETs and the solving domains. Among all, two commercial packages have been used to design the different structures throughout this dissertation. They are called computer simulation technique Microwave Studio (CST MWS) and Ansoft High Frequency Structure Simulator (HFSS) [73, 74]. While CST MWS is based on time domain finite integration technique (FIT), the Ansoft HFSS is based on frequency domain finite element method (FEM). This gives the opportunity to cross check the results with both techniques. Hence, all the structures are measured afterwards; the reliability of the programs has been evaluated.

Generally, computer simulation programs give results anyway. Therefore, using the aforementioned packages without a pre-knowledge of the methods could be misleading. So, it is important to understand the methods, which are used by the packages. Therefore, the basics of the FIT and FEM are summarized in this section. Moreover, some methods are more suitable for some structures than others depending on the sharpness of their responses. Fig. 3.3 shows the different techniques versus the quality (Q-) factor. While the time domain techniques are suited for broadband responses, the eigen mode analysis represents an efficient tool for highly resonant structures leaving the frequency domain analysis in the middle. It may be noted here that a detailed description of these methods is beyond the scope of this chapter. Further details and an elaborated description of these methods can be found in [75, 76, 77].

3.1.1 CST MWS - Finite Integration Technique

CST MWS is a fully featured software package for electromagnetic analysis and design in the high frequency range. It is based on finite integration technique (FIT) to compute the electromagnetic behavior of microwave components. The finite integration technique is based on a consistent discrete representation of Maxwell's equations on grids. In this method the integral form of the Maxwell's equations are used for discretization. This method can be considered as a generalization of the finite-difference time-domain (FDTD) method [75] sharing its advantages and disadvantages. In this technique, the six components of electric field strength and magnetic flux density are computed on a grid mesh system. Some steps are required to perform this. The first step is to model the required component(s); then a fully automatic meshing routine is applied before the simulation is started. Moreover, defining a finite calculation domain out of the infinite size of the open boundary is essential in order to compute the electromagnetic fields with the help of a computer with finite memory. The considered application problem is enclosed by this finite calculation domain. The Maxwell's equations in integral form, which are the governing equations used for the field computation in FIT are given by:

$$\oint \mathbf{E}ds = -\int \frac{\partial \mathbf{B}}{\partial t}dA \tag{3.1}$$

$$\oint \mathbf{H}ds = \int (\frac{\partial \mathbf{D}}{\partial t} + \mathbf{J})dA \tag{3.2}$$

$$\oint \mathbf{D}dA = \int \rho dV \tag{3.3}$$

$$\oint \mathbf{B}dA = 0 \tag{3.4}$$

Here, $\mathbf{E}$ and $\mathbf{H}$ are the electric and magnetic field vectors, $\mathbf{D}$ and $\mathbf{B}$ are the electric and magnetic flux density vectors, ρ is the electric charge density and $\mathbf{J}$ is the current density, respectively. The electromagnetic fields which are related to each other according to equations 3.1-3.4 are to be computed within the computational domain defined earlier. In the next step, the computational domain is decomposed into a finite number of volumetric cells, termed as FIT cells. The FIT cells are formulated in such a way that they exactly fit to each other. This decomposition scheme yields the finite integration grid which serves as the computational grid. However, there are different methods to do that. Fig. 3.4 represents the three best known methods for meshing. While Fig. 3.4(a) shows the staircase meshing, Fig. 3.4(b) and 3.4(c) demonstrate the tetrahedral and the perfect boundary approximation (PBA) methods, respectively.

Stability is another important feature for such a technique. For regular equidistant coordinate grids filled with homogeneous materials, this condition relates the largest time step Δt allowed with the smallest mesh steps in the three spatial directions as,

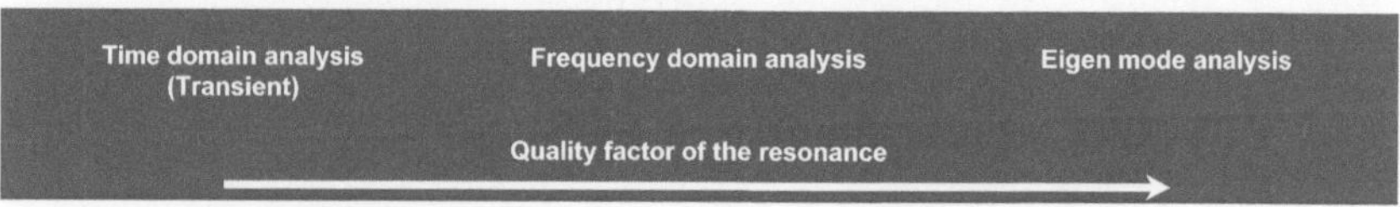

Figure 3.3: Different MWS CST solvers versus quality factor.

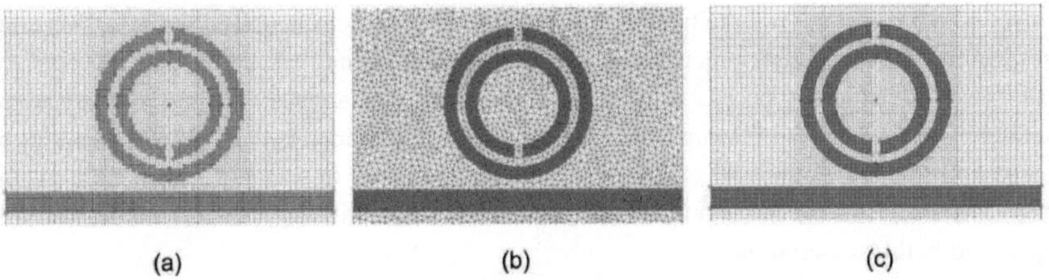

Figure 3.4: Different CST meshing methods: (a) staircase (b) tetrahedral (c) perfect boundary approximation.

$$\Delta t \leq \frac{\sqrt{\mu\epsilon}}{\sqrt{(1/\Delta x)^2 + (1/\Delta y)^2 + (1/\Delta z)^2}} \tag{3.5}$$

Here, Δx, Δy and Δz are step sizes in the x, y and z direction respectively, whereas ϵ and μ denote the permittivity and permeability of the material in the cell. The stability condition in Eq. 3.5 basically enforces the causality condition by keeping the maximum allowed time step value equal to the time taken by electromagnetic waves to pass the smallest cell. It can be seen from Eq. 3.5 that in order to obtain a stable scheme for a problem with very fine mesh size, the allowed time step should be very small leading to a large simulation time. Moreover, Fig. 3.5 shows the percentage error of the calculation versus the number of steps per wavelength which reveals an exponential decrease.

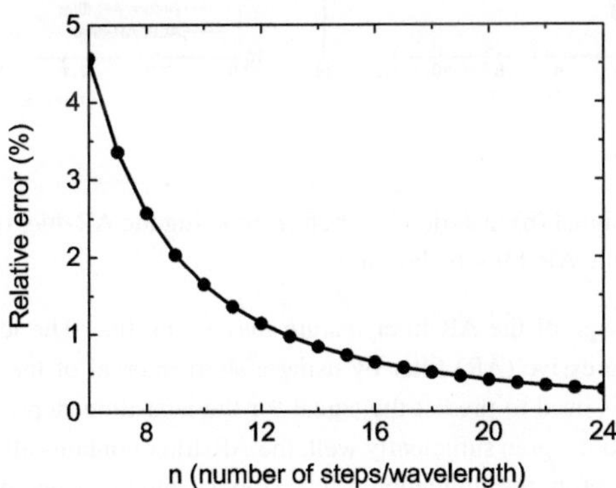

Figure 3.5: Percentage error versus number of steps/wavelength.

The most flexible tool in CST MWS is the transient solver, which can obtain the entire broadband frequency behavior of the simulated device from only one calculation run. However, efficient filter design often requires the direct calculation of the operating modes in the filter rather than an S-parameters simulation. For these cases, CST MWS also features an eigenmode solver which calculates a finite number of modes in closed electromagnetic devices. When investigating highly resonant structures such as narrow bandwidth filters, a time domain approach may become inefficient, because of the slowly decaying time signals. The usage of advanced auto-regressive signal processing techniques (AR-filters) provided by CST MWS allows us to speed up these simulations by orders of magnitude compared to standard time domain methods. The transient and the eigenmode solver have been extensively used for the simulations presented in this dissertation.

The performance of the transient solver can be significantly reduced when strongly resonating devices are simulated. The main reason for this is that the calculation of the S-parameters by using a Fourier Transform requires the time signals to have sufficiently decayed to zero (see Fig. 3.6(a), otherwise a so-called truncation error will occur. Especially for highly resonating devices, a ringing in the time signals may occur such that the signals decay very slowly and therefore require long simulation times for accurate Fourier Transforms. If S-parameters calculation is performed by using time signals which still exhibit some oscillation, a ripple will appear on the S-parameters as shown by dotted curve in Fig. 3.6(b) below:

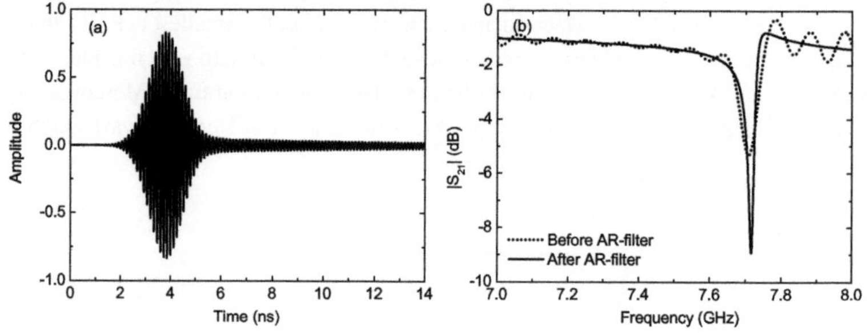

Figure 3.6: (a) Time signal (b) Insertion loss before applying the AR-filer (dotted line) and after applying the AR-filter (solid line).

Otherwise, the usage of the AR-filter feature may be useful. The idea of this technique is to train an auto-regressive (AR) filter by using a short interval of the time signals. Afterwards, the AR-filter is used to predict the signal for the next time steps. Once the prediction and the actual simulation agree sufficiently well, the AR-filter contains all relevant information about the device. Therefore the simulation can be stopped and the S-parameters can be derived mathematically from the AR-filter representation as shown by solid curve in Fig. 3.6(b).

3.1.2 Ansoft HFSS - Finite Element Method

HFSS is a full-wave simulation software for the electromagnetic problems of high-frequency components [74]. Ansoft HFSS utilizes a 3D full-wave finite element method (FEM) to calculate the electrical behavior of high-frequency components. It is an interactive software package for calculating the electromagnetic behavior of a structure. The software also includes post-processing commands for analyzing the electromagnetic behavior of a structure in more detail.

Generally, the FEM formulation is more complex than the FDTD method. The FEM was developed from the need to solve complex elasticity, structural analysis problems in civil engineering. The generality of this method makes it possible to build general purpose computer programs which can be applied to solve different kind of problems. There are several advantages associated with the FEM. With this method, complex geometries can be straightforwardly and conveniently modeled. With FEM, each mesh element can be defined independently. Thus, a large number of fine mesh elements can be used in regions with complex geometry and fewer, larger elements can be used in relatively simpler or open regions. This feature allows an efficient modeling of complex and large structures. Being a general method, this method has the potential to couple many problems like the mechanical, thermal and electromagnetic problems together.

In order to generate an electromagnetic field solution, HFSS employs the finite element method. In general, the finite element method divides the full problem space into thousands of smaller regions and represents the field in each sub-region (elements) with a local function. In HFSS, the geometric model is automatically divided into a large number of tetrahedral, where a single tetrahedron is a four-sided pyramid. This collection of tetrahedral is called the finite element mesh. The smaller the elements are, the more accurate is the solution. There is a trade-off between size of the element and computing resource available. For a large structure, finer mesh means a large number of elements which requires a large amount of computer memory and processing resources. Hence, the mesh size should be carefully chosen. A coarse mesh size leads to better accuracy of the solution and too fine mesh size overwhelms the available computing resources. It is advisable to specify the mesh size smaller than a quarter wavelength of the wave of excitation.

To optimize the mesh size, HFSS uses an adaptive iterative meshing method at the solution frequency. In the first iteration HFSS generates a solution based on an initial mesh. It checks the solution accuracy to a user defined value. If error tolerance is not accomplished, it refines the mesh in critical areas and generates a new solution. When the error tolerance converges to a user defined accuracy, HFSS stops the loop. Also, the maximum number of iterations can be set to a predefined number. When this number is reached even though the error has not converged, HFSS stops the loop with a warning indicating error not converged. The above mesh refinement happens at the solution frequency selected.

3.1.3 CST MWS versus Ansoft HFSS

CST MWS solves for transient (time-domain) E-field directly. That is, it directly solves for electric field in a 3D space as a function of time. It then indirectly, through Maxwell's equations, solves for H-field as a function of time. Once **E** and **H** are known, many other related parameters (such as current distribution), can be computed indirectly. Frequency domain parameters are indirectly obtained through Fourier transformation.

In contrast, HFSS solves for E-field as a function of frequency. **H**-field is indirectly obtained through Maxwell's equations. Other related parameters are indirectly computed just as in MWS. Time-domain parameters are indirectly obtained through inverse Fourier transformation.

Moreover, anything computed directly is "potentially" more accurate than what is computed indirectly. Therefore, transient parameters computed by MWS are "potentially" more accurate than transient parameters computed by HFSS. Likewise, frequency-domain parameters computed by HFSS are "potentially" more accurate than frequency-domain parameters computed by MWS. Certain Fourier transformation conditions should be met to obtain accurate results when the parameters are changed from one domain to the other. In most cases the conditions are hard to meet thereby limiting the accuracy.

3.2 Structures Preparation Steps

The next step after the design optimization is to print the mask and to fabricate the structures. Fig. 3.7 shows the flow chart of the different steps. Two kinds of substrates have been used in this dissertation. The first one is covered with a photoresisit and the second one is not covered with the photoresisit. The photolithography process has been used to prepare the structures. Photolithography is a process used in the fabrication to selectively remove unwanted areas of the metal film. It uses light to transfer a geometric pattern from a photomask to a light-sensitive chemical photoresist on the substrate.

The photoresist is dispensed onto the wafer, and the wafer is spun rapidly to produce a uniformly thick layer. The spin coating typically produces a layer between 0.5 and 2.5 μm thick. The spin coating process results in a uniform thin layer, usually with uniformity within 5 to

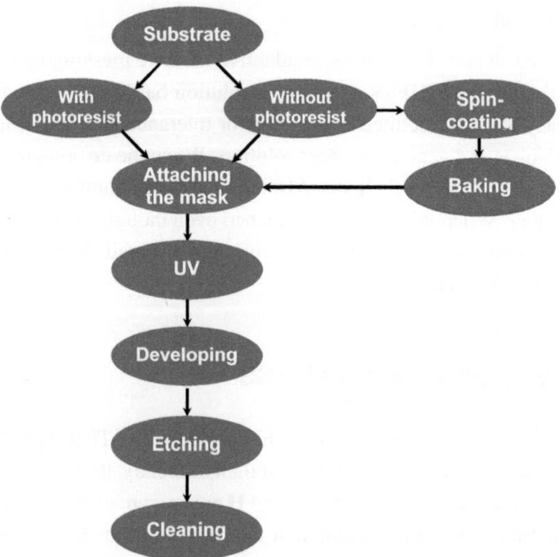

Figure 3.7: Flow chart of the structure preperation steps.

10 nanometers. After pre-baking, the photoresist is exposed to a pattern of intense light. Photolithography typically uses ultraviolet (UV) light. This chemical change allows some of the photoresist to be removed by a special solution, called a "developer" by analogy with photographic developer. The resulting wafer is then "hard-baked", typically at 120 to 180 °C [78] for 1 to 2 minutes. The hard bake indurates the remaining photoresist, to make a more durable protecting layer. Etching is a subtractive process used to remove unwanted parts in layer patterning. In etching, a liquid ("wet") chemical agent removes the uppermost layer of the substrate in the areas that are not protected by photoresist. After a photoresist is no longer needed, it must be removed from the substrate. This usually requires a liquid "resist stripper", which chemically alters the resist so that it no longer sticks to the substrate.

3.3 Measurements

The measurements have been performed using a vector network analyzer HP E8361A from Agilent. It is an instrument which measures the complex transmission and reflection characteristics of two-port devices in the frequency domain. It does this by sampling the incident signal, separating the transmitted and reflected waves, and then performing ratios that are directly related to the reflection and transmission coefficients of the two-port (S-parameters). Frequency is swept to rapidly obtain amplitude and phase information over a band of frequencies of interest. The S-parameters S11 and S21 can be interpreted as the input reflection coefficient and insertion loss coefficient under the conditions that the source and load impedances represent perfect Z_o (e.g. 50 Ohm) conditions. User calibration is necessary to correct for cable losses, to correct for non-Z_o conditions, and to establish phase reference planes.

3.3.1 Designing the TRL Calibration Kits

Measurements at microwave frequencies are usually made at some reference plane, physically removed from the device-under-test (DUT), as illustrated in Fig. 3.8. The error boxes represent necessary test adapters that allow the measurement equipment to interface with the DUT. The purpose of network analyzer calibration is to determine the values of all the S-parameters in the error boxes at each frequency of interest.

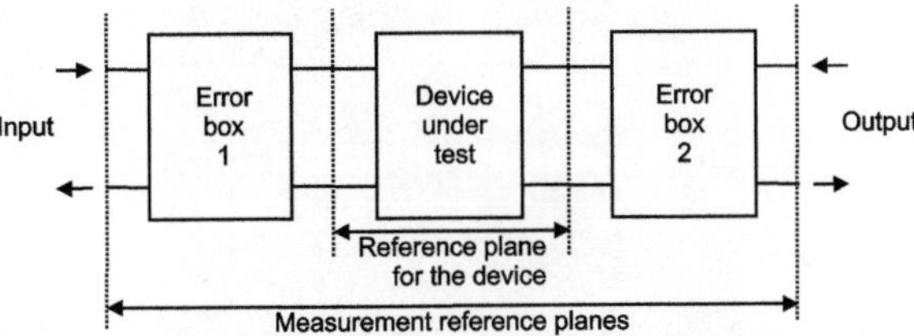

Figure 3.8: Schematic representation of measurement at reference planes physically removed from the DUT.

A well known calibration method utilizing Thru-Reflect-Line (TRL) is generally used. Many names have been given for this generalized calibration system such as Thru-Short-Delay and Thru-Reflect-Match. These techniques involve the same basic approach, only with distinct loads, as suggested by the name.

One is the thru-reflect-line, or TRL, which can be used to calibrate the network analyzer. Like other calibration methods, the TRL introduces a 12-term error correction vector for each frequency point. To calculate these terms, standards for which S-parameters are known must be measured. These standards include a short order open, a thru, and a delay line with known electrical delay. The TRL calibration corrects phase and magnitude errors introduced by the sliding of reference planes and the insertion loss of cables, fixtures and connectors. The three connections for TRL calibration of microstrip are:

1. Thru: connecting port 1 to port 2 directly.

2. Reflect: Terminating a microstrip connected to each port with an open or short load that produces a large reflection.

3. Line: Connecting the two ports together thru a line approximately 1/4 longer than Thru.

3.3.2 CPW Structures Measurements

In order to avoid any undesirable effects due to connectors and so on, Wiltron 3680 test fixture (see Fig. 3.9) is used along with the VNA to measure all the CPW structures. In fact, the most critical part of any substrate measurement system is the launching fixture. It must be simple and flexible, and most of all provide accurate and repeatable measurements. The fixture consists of a fixed connector and a movable connector that can be positioned for substrates up to 5 cm long. The substrate is held in place between spring loaded jaws. This allows the fixture to accommodate different devices without requiring a custom center section for each different length. The unique jaw action ensures solid, repeatable electrical contact. The jaw tension is defined by the force of a spring, independent of human judgment errors. This means the tension will always be

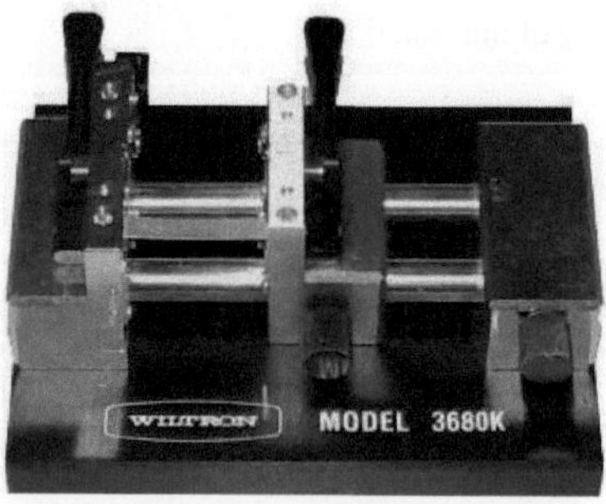

Figure 3.9: Wiltron 3680 text fixture.

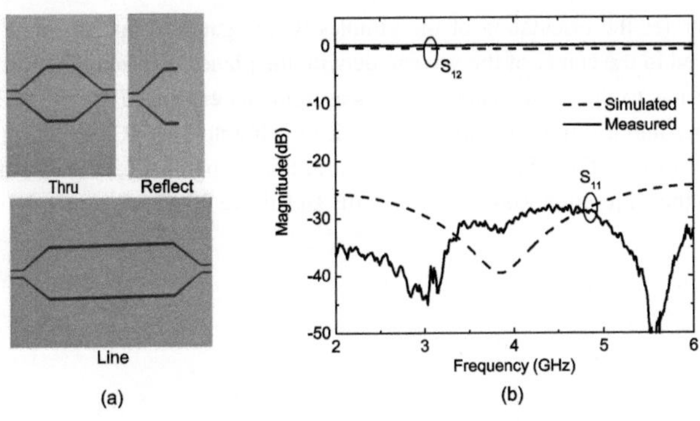

Figure 3.10: (a) CPW PCB calibration kit (b) S-Parameters of the thru.

the same, providing more repeatable measurements. Complete substrate measurement systems comprising of a test fixture, a vector network analyzer, and a TRL calibration kit can fulfill microstrip or CPW test needs. Fig. 3.10(a) and (b) shows the TRL kit for one of the substrates using the CPW and its S-parameters. The return loss is better than 26 dB for the band of interest and the insertion loss ripple is less than 0.02 dB

3.3.3 Metasurfaces Structures Measurements

In this work, instead of a conventional FSS free-space setup, we employ a single mode rectangular waveguide with a single unit cell structure positioned inside. Thus, the experiment is more robust, as it is shielded against environmental influences, and highly reproducible. Furthermore, the waveguide confines the fields in a small region around the unit cell, significantly increasing the effective scattering cross-section of the resonator. Fig. 3.11(a) shows the general

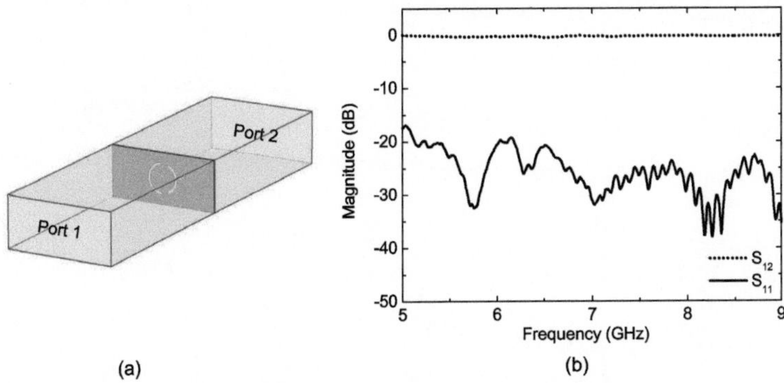

Figure 3.11: (a) General setup: a single unit cell is loaded inside the rectangular waveguide (b) S-Parameters of the calibration kit thru.

setup and illustrates the orientation of the samples with regards to the incoming wave. The unit cell is placed in the center of the waveguide with the plane containing the resonator structures perpendicular to the wave vector. The measurements employed an HP E8361A vector network analyzer and a thru-reflect-line (TRL) kit to calibrate the WG14 rectangular waveguide. Fig. 3.11(b) shows the S-Parameters of the thru. For the band of interest, the insertion and return losses in the calibration were below 0.05 dB and above 20 dB, respectively.

4 CPW Filters Based on Novel Metamaterial Resonators

This chapter reviews the novel planar metamaterial resonator concepts which have been proposed throughout this work for filter applications. It is divided into two main parts. While the first part deals with the CPW bandstop filters consisting of a single metal layer, the second part introduces compact bandpass filters based on metamaterial resonators. The chapter ends with a short summary.

The first part begins by introducing the complementary split ring resonators (CSRRs). Unlike the structures introduced in Sec. 2.4.2, which consist of two metal layers, the CSRRs have been engaged with the CPW in the same layer to maintain the advantages of the single layer CPW. Moreover, these resonators provide a distinct stopband characteristic, which can be adjusted by adding slots in the proximity of the CSRRs. Modifying the slot lengths offers great design flexibility. However, the stopband response suffers from a spurious rejection band close to the main resonance, which has its origin from the difference in electrical length of the inner and the outer resonator arm. This issue leads to the u-shaped CSRR (CUSR), which cancels the spurious resonance by equalizing the electrical length of both resonator arms. To further miniaturize the CSRRs, rectangular multiple turn complementary spiral resonators (CSRs) are proposed. It enables extremely small electrical footprint filters as each turn elongates the effective resonator length, thus lowering the resonance frequency. A very small electrical size of $\lambda_g/50$ is achieved.

The second part starts with the demonstration of the effect of slots inserted in the top layer of the CPW and in direct proximity to split ring resonators. The slots enable a full control of the bandwidth by varying their lengths. This concept could also be applied to other resonator designs. Another concept is demonstrated to achieve very compact BPFs is by combining conventional SRRs with CSRRs. In addition to the circular structures, a rectangular spiral design (RSRs) will be investigated, which exhibits an even higher miniaturization potential, which could be very useful, e.g. in front-end filter designs.

4.1 Introduction

As the demand for microwave and millimeter-wave circuits increases, innovative circuit architectures are required to suppress frequency parasitics and harmonics. To overcome the limitations of traditional techniques, electromagnetic bandgap (EBG) structures have been proposed [79, 80]. Yet, for some applications EBGs can be relatively large because they rely on interference. Hence, the unit cell has dimensions of at least a quarter of the design wavelength and quite a few unit cells are needed to provide significant rejection. As an alternative to EBGs, a planar transmission line can be loaded with split-ring resonators (SRRs). SRRs were originally proposed by Pendry [5]. Arranged periodically, they can form an effective medium with a negative magnetic permeability in the vicinity of their resonant frequency. The electrical dimensions of SRRs are small in comparison to the design wavelength because they are sub-wavelength structures.

Moreover, many modern microwave applications, such as automotive radar and broadband wireless communication, rely on high frequency bandpass filters (BPFs). As the operation frequencies of microwave devices continue to rise and ever tighter circuit integration is required, many conventional filter designs reach their technologically defined limits. To satisfy the increasing demand for high performance, small outline and low cost millimeter wave BPFs, alternative concepts have to be explored [81, 82, 83]. Coplanar waveguide (CPW) technology has gained considerable popularity in the development of microwave circuits, as CPW structures provide many conceptual advantages over conventional microstrip designs. These advantages include lower dispersion, easy mounting of lumped elements, simple series and shunt connections as well as insensitivity to substrate thickness [45, 84, 85]. Designing BPFs with a wide stopband and a sharp rejection is a challenging task, as it requires a multi-stage approach. Due to the resulting high order transfer function, conventional filters exhibit a poor out-of-band response [86]. In this chapter, some novel planar metamaterial resonator concepts will be reviewed.

4.2 CPW Compact Bandstop Filters-Recent Advances

4.2.1 Single Layer CPW Filter Synthesis

The CPW transmission line, which consists of one metal layer, holds potential advantages for microwave circuits as discussed earlier. Hence, applying the metamaterial SRR principle and maintaining these advantages is highly desirable. There have been some attempts to achieve that as discussed in Sec.2.4.2. Unfortunately, the overall performance is quite poor for those structures.

In the following, we will show step by step how to obtain a CPW one metal layer bandstop filter. First of all, let's consider a very simple structure as shown in Fig. 4.1(a). It consists of a microstrip line loaded with two SRRs on the top of the substrate with a ground plane on the bottom side. Applying Babinet's principle to this structure leads to a slot line loaded with CSRRs as depicted in Fig. 4.1(b). The CPW requires another slot line as in Fig. 4.1(c). The unit cell of this structure can be modeled (Fig. 4.1(d)) as a CPW loaded with CSRRs [87]. As mentioned earlier, many cells are required to achieve acceptable performance [88]. Fig. 4.2 shows the transmission response of four structures having one, two, three, and four unit cells, respectively. It is noticed that many cells are required to achieve a good rejection level. This is in agreement with the results presented in Ref. [88]. Studying the field distribution in the CPW (Fig. 2.13), shows that the electric field is enhanced close to the gaps of the CPW. This

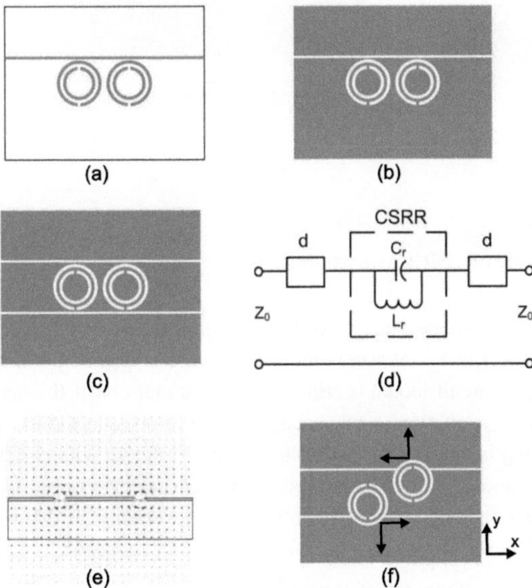

Figure 4.1: (a) SRRs loading a microstrip line (b) CSRRs loading a slot line (c) CSRRs loading the center conductor of a CPW (d) the equivalent circuit model of a unit cell of the structure in (c) (e) E-field distribution in the CPW (f) the direction in which the CSRRs are moved.

inspired us to move the two CSRRs in the opposite y-directions (as indicated by the vertical arrows in Fig. 4.1(e)) until their centers coincide with the gaps of the CPW. In addition, the CSRRs are moved towards each other in x-directions (as indicated by the horizontal arrows in

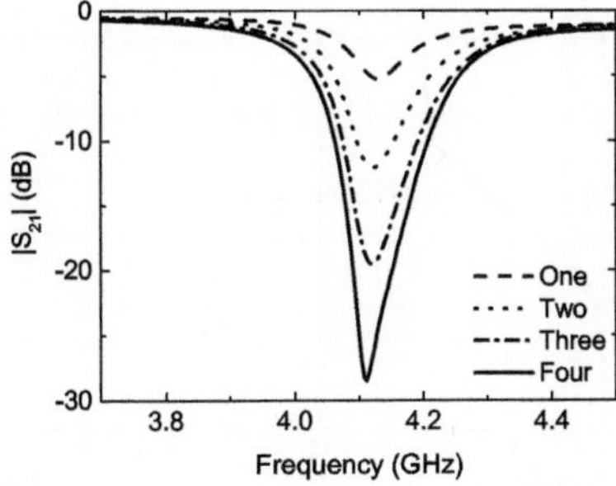

Figure 4.2: The transmission response of one, two, three, and four unit cells etched into the center conductor of the CPW.

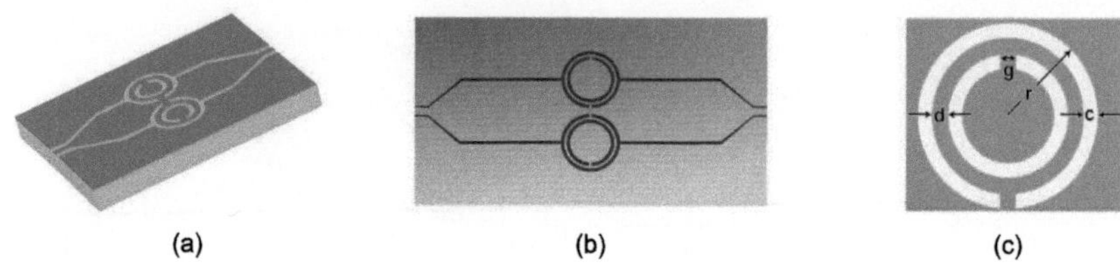

(a)   (b)   (c)

Figure 4.3: (a) 3D layout of CPW/CSRR structure (b) top-view of the fabricated structure (c) schematic of CSRR with its dimensions.

Fig. 4.1(f)) until they hold the same x-position. For this geometry the excitation of the CSRRs is enhanced. Moreover, the total size required to achieve a specific performance will be much less. The equivalent circuit model is still valid for the unit cell if the magnetic wall concept is considered. As the excitation is enhanced, this should be reflected in the parameters of the model. The resulting layout is a compact planar metal structure with complementary split ring resonators, which should exhibits a high rejection stopband. Moreover, it is metalized only in one layer and, hence, is more compact than the structure studied by Martin et al. [48]. With a single pair of CSRRs we obtain a considerably higher suppression than reported in Ref. [48].

Fig. 4.3(a) shows a 3D layout of the integrated CPW/CSRRs structure. First, we simulate a single unit cell with periodic boundary conditions in propagation direction. The simulation is performed using the Eigenmode solver of CST Microwave Studio [73]. While the fabricated structure is shown in Fig. 4.3(b), the dimensions of the unit cell are shown in Fig. 4.3(c). They are chosen such that the structure operates in the C-band. The CSRR has an external radius of $r = 3.6$ mm, a width of $c = 0.27$ mm, a separation of $d = 0.43$ mm, and a length of the "metallic bridge" of $g = 0.43$ mm. The dispersion is calculated by varying the phase shift in propagation direction between 0° and 180°. We observe a forbidden gap between 4.2 GHz and 5.6 GHz as shown in Fig. 4.4(a). Hence, we shall expect a stopband behavior in the transmission magnitude with a center frequency around 4.9 GHz. This bandgap is provided by the complementary rings

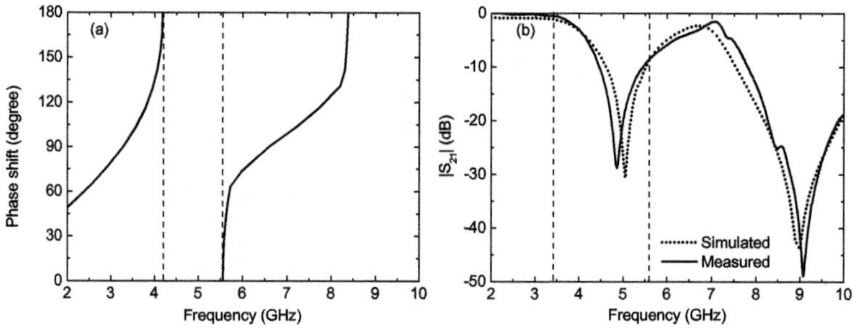

Figure 4.4: (a) Dispersion analysis (b) simulated and measured insertion loss.

with an effective negative dielectric permittivity. Two CPW tapers are added at the edges of the structure to exclude measurement errors due to soldering connectors to the devices. The taper was verified through simulation and experiment to provide maximum matching between the two sides of the CPW. A standard mask/photoetching technique is used to fabricate the structures using an FR-4 substrate (dielectric constant $\epsilon_r = 4$, loss tangent $\tan \delta = 0.02$, thickness $h = 0.5$ mm).

A commercial software (Ansoft HFSS [74], a 3D full-wave solver based on the finite element method with adaptive iterative meshing) is used to simulate the transmission through the structures. A very high mesh resolution is utilized by specifying the maximum change in the S-parameters between two successive iterations to be 0.3%. Moreover, a fine discrete sweep is performed with a 0.02 GHz step size for the band between 4.5-5.5 GHz, where we expect rapid changes in the transmission. Simulated and measured magnitude for the insertion losses are shown Fig. 4.4(b) in dotted and solid lines, respectively. The insertion loss for the CSRR structure shows the expected stopband behavior close to 5 GHz. If -10 dB is taken as reference level the stopband ranges from 4.6 GHz to 5.5 GHz in very good agreement with the dispersion analysis. The S-parameters of the fabricated structures are measured between 2 and 10 GHz using an HP E8361A vector network analyzer (VNA) with a microstrip test fixture (Wiltron 3680). A thru-short-line (TRL) calibration was performed for the CPW. The return loss is better than 26 dB for the band of interest. The measured S-parameters are in good agreement with the simulated HFSS results and confirm the expectation of the dispersion analysis carried out with CST MWS. The small shifts in the resonance frequency are due to inhomogeneities in the ring dimensions. Beneath the low-frequency limit of the artificial band gap, the structure exhibits excellent matching without any significant insertion loss. In the upper passband a good performance up to 7 GHz is achieved, but for higher frequencies a spurious resonance leads to an undesired dip in the transmission response at 9.1 GHz. The origin of this dip and counter-measures to remove it are discussed in the section following the next one. Another remarkable aspect about the CSRRs is the high rejection level in the forbidden band of nearly 30 dB, which is at least twice as high as the suppression of the SRR-based structures introduced in [48]. Thus, CSRRs provide an effective way to eliminate frequency parasitics in CPW structures. By cascading multiple CSRR unit cells even higher suppression levels can be achieved. To further visualize the effectiveness of CSRRs, the electric field was simulated at the passband (2 GHz) and at the stopband (5 GHz) as illustrated in Fig. 4.5. Most of the signal gets through the strucutre at the passband frequency, while the CSRRs stops almost all the signal.

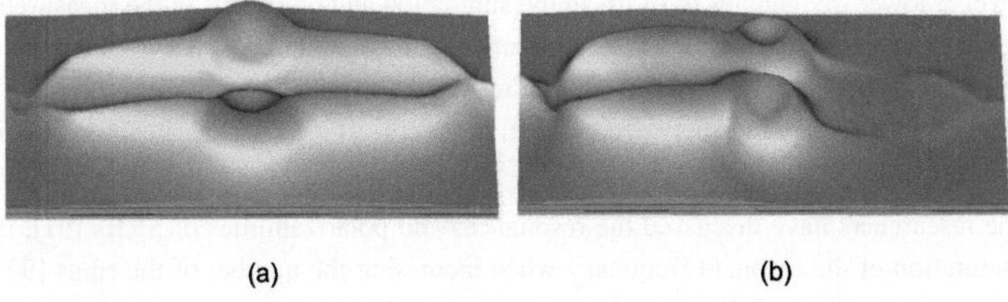

(a)   (b)

Figure 4.5: 3D electric field distribution at 2 GHz (a) and 5 GHz (b).

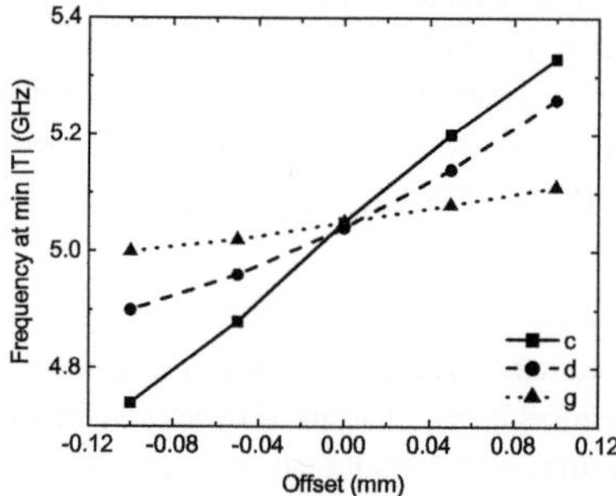

Figure 4.6: Resonance frequency of the CSRR-slots structure as a function of offset in the CSRR dimensions.

Further investigation has been done to examine the effect of different geometrical dimensions on the resonance frequency of the structure. The resonance frequency drops considerably if the ring radius is increased. The same effect has been found for the SRR structure [89]. Fig. 4.6 shows results obtained from a large set of simulations. It is found that the resonance frequency increases with the width c and the separation d between the rings. This dependence is shown as squares and circles, respectively in Fig. 4.6. For the SRR the opposite (dual) effect has been observed [89].

4.2.2 CPW-SRRs and CPW-CSRRs Comparison

We shall presenting the simulated and measured results for the structures with a single and three SRRs cells. The dispersion analysis (will be shown in chapter 5) shows that there is an unexpected narrowband mode around 5.5 GHz which does not appear in the cases of outer or inner rings only. This is a result of having two resonators with close modes. The same effect has been observed for metamaterials bandpass filters [90]. Fig. 4.7(a) shows simulated (dashed line) and measured (continuous line) magnitudes for the insertion and return losses for a single unit cell of SRRs. A notch in the transmission response is observed in the vicinity of the magnetic resonance frequency. The suppression level is about 15 dB. The stop band is rather narrow because the negative permeability occurs in a narrow bandwidth. The return loss (S11) level at lower frequencies is 14 dB in the simulation and only 9 dB in the measurements. However, the measured data obtained experimentally is in good agreement with the simulation. A second mode is observed around 5.5 GHz. It results from the coupling of the outer and inner rings. The small discrepancies between simulation and measurements are attributed mainly to the inhomogeneities in the SRRs.

Some researchers have discussed the resonances and polarizabilities of SRRs [91], the inherent saturation of the resonant frequency when increasing the number of the rings [92], and the miniaturization of the CSRRs loaded microstrip lines [93]. We carried out an extensive study for this kind of filter design. We simulated structures which have only an outer ring,

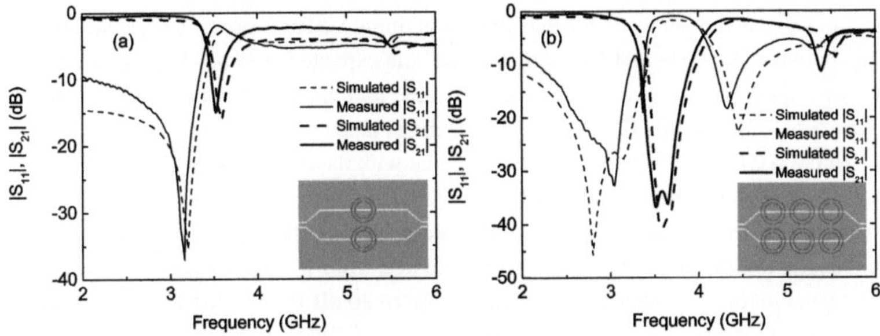

Figure 4.7: (a) Simulated (dashed line) and measured (solid line) insertion and return losses for a CPW loaded with a single pair of SRRs. (b) Simulated (dashed line) and measured (solid line) insertion and return losses for a CPW loaded with three SRR pairs.

only an inner ring, or both rings. We found magnetic resonance frequencies of 3.92 GHz, 4.9 GHz, and 3.6 GHz, respectively. The rejection level at the resonance frequency was almost the same. This means that one could use only a single ring instead of a concentric double SRR. Yet, miniaturization is supported by using two concentric rings. Our studies show that the two rings couple to each other leading to a new resonance frequency which is lower than that of both of the two individual rings. Of course, this lower frequency could also be obtained by using one single ring, but this would have to have larger radius. This finding inspired us to investigate structures with an even higher number of concentric rings. However, the resonance frequency deceases only another 2% when going from two to four concentric rings (two rings were added in the middle of the two ring structure to keep the outer SRR dimensions the same).

Hence, only a very small spectral shift would be observed if further rings are added. This can be understood intuitively as adding concentric rings inside the first two rings increases the metallic area in the middle of the structure. There is a saturation in the total resonator parameters, so the shift in the resonance frequency saturates [92]. Moreover, adding more rings with the dimensions given above ($r = 3.6$ mm, $c = 0.27$ mm, and $d = 0.43$ mm) will deteriorate the field exciting the rings [93]. The inner rings will be excited only with a very little amount of the magnetic field. Therefore, it is logical that the shift in the resonance frequency saturates. However, reducing the distance between the rings will be an avenue to increase the coupling and, hence, to decrease the resonance frequency. SRRs are considered as being lumped or quasilumped elements in [91]. Their electrical size is $\lambda_g/8$ according to the outer radius of the SRRs and their resonance frequency, where λ_g is the guide wavelength. While it is $\lambda_g/6$ in the case of CSRRs. However, for a fair comparison one should take into account two structures with the same performance and compare them. Moreover, the structure with CSRRs is only one metal layer that might be desirable in many applications. Fig. 4.7(b) shows the simulated and measured insertion and return losses for three SRR cells. One notices a significant improvement in the suppression level. In this case it is almost 40 dB. Furthermore, the increase in the pair number leads to a sharper rejection band. The measurements are in very good agreement with the simulations. For frequencies above 4 GHz the return loss is higher than that of the structure with the single pair.

Previous dispersion analysis for the CPW containing CSRRs shows that the structure has a bandgap between 4.2 GHz and 5.6 GHz [94], i.e. the expected stop-band resonance frequency is close to 5 GHz. Fig. 4.8(a) depicts the simulated and measured insertion and return losses for a structure with a single CSRR cell. If -10 dB is taken as a reference level, the stop-band ranges from 4.6 GHz to 5.5 GHz in good agreement with the dispersion analysis. The simulated results and the calculations of the dispersion characteristics are confirmed by the measured S-parameters of the fabricated structure. However, outside the frequency gap, the structures exhibit excellent matching with insignificant insertion loss. It is remarkable that the measured rejection level in the forbidden band, which is close to 30 dB for the CSRR structure, is at least twice that of the structure with an SRR. Multiple pairs of SRRs will be necessary to obtain this suppression level. This means that this structure with only one cell and only one metal layer can achieve the same behavior as a double layer structure with many SRR cells. The discrepancy between the resonance frequencies of SRR and CSRR arises mainly from the effect of the dielectric substrate affecting the resonance frequencies in a different way. This is in agreement with the results in Ref. [49].

Hence, it is clear that CSRRs provide an effective way to eliminate frequency parasitics in CPW structures. Moreover, the return loss level at lower frequencies is about 20 dB, which is again better than that of the structure containing SRRs. In addition, having only one metal layer simplifies the fabrication process considerably. The simulated and measured insertion and return losses for a structure with three CSRR cells are shown in Fig. 4.8(b). The rejection level is higher and the bandwidth is increased as compared to the case of only one CSRR cell. The 10 dB bandwidth ranges from 4 GHz to 5.65 GHz and is in better agreement with the dispersion analysis than that observed for one CSRR cell. This is indeed logical and understandable, since the dispersion analysis is carried out for a structure which has an infinite number of cells. A remarkable improvement in the suppression level is observed. It is almost 50 dB in this case. The measurements are again in very good agreement with the simulations.

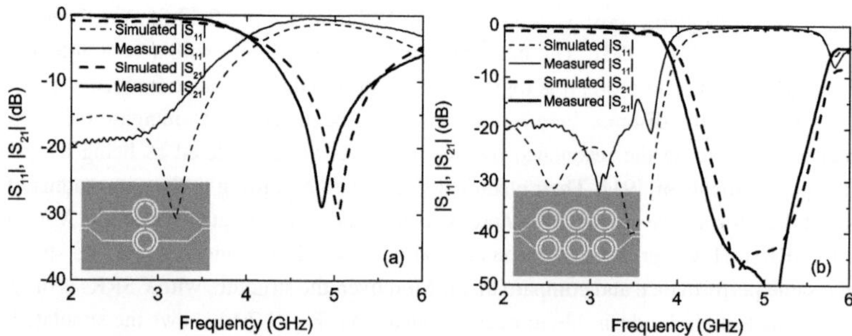

Figure 4.8: (a) Simulated (dashed line) and measured (solid line) insertion and return losses for a CPW incorporating a single pair of CSRRs. (b) Simulated (dashed line) and measured (solid line) insertion and return losses for a CPW incorporating three CSRR pairs.

4.2.3 Spurious Elimination Using Miniaturized CUSRs

The presence of spurious bands is a fundamental limitation of microwave filters implemented by means of distributed elements. These undesired frequency bands can seriously degrade filter performance and may be critical in certain applications that require huge rejection bandwidths. Unfortunately, for most filter implementations, the first spurious band is relatively close to the frequency region of interest. In the last section, we discussed the CPW-integrated CSRRs for use as a single metallization layer, high rejection stopband filter. Yet, the insertion loss of the CPW-CSRR suffers from a strong spurious stopband located on the high frequency side very close to the first stopband. This results from a phenomenon which we call differentiating resonances [90]. This inspired us to propose a complementary u-shaped split resonator (CUSR). The physical length of the u-shaped outer and inner rectangular resonator is the same. This fact should prevent the second resonance of the CSRR and therefore should lead to a smooth performance in the insertion loss. Moreover, we investigate the effect of the relevant CSRR and CUSR dimensions on their resonance frequencies in a comparative study.

The CPW-CSRR structure was initially proposed by the author in 2007 [94]. Fig. 4.9 shows the layout of the tapered CPW incorporated with one CSRR or CUSR cell, respectively. The resonators are symmetrically etched into the top layer. The lattice constant a is 9.6 mm while the center conductor and slot width are 9.15 mm and 0.45 mm, respectively. At the end of the structure the dimensions of the center conductor and slot width are 1 mm and 0.14 mm. These lateral dimensions have been chosen to obtain a characteristic impedance for the host line of Z_o = 50 Ohm. The CPW tapers are added at the edges of the structure to exclude measurement errors due to soldering connectors to the devices. The taper was verified through simulation to give maximum matching between the two sides of the CPW. An in-house standard mask/photoetching technique is used to fabricate the structures on FR-4 substrate (dielectric constant ϵ_r = 4, loss tangent tan δ = 0.02, thickness h = 0.5 mm, and double sided copper clad

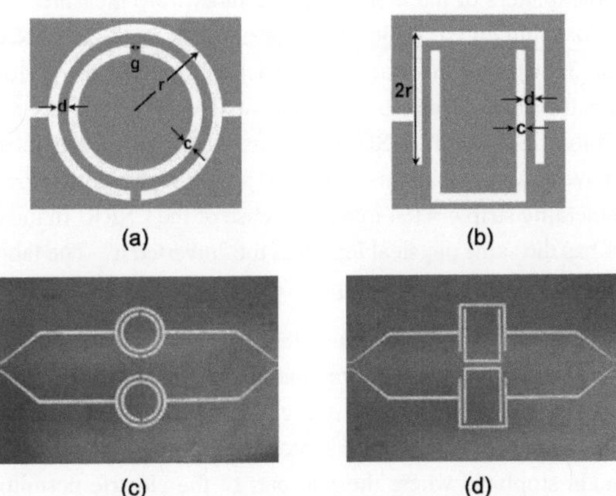

Figure 4.9: (a) Schematic of CSRR and CUSR (b) with their dimensions and layout of the fabricated CPW/CSRR structure (c) and CPW/CUSR structure(d).

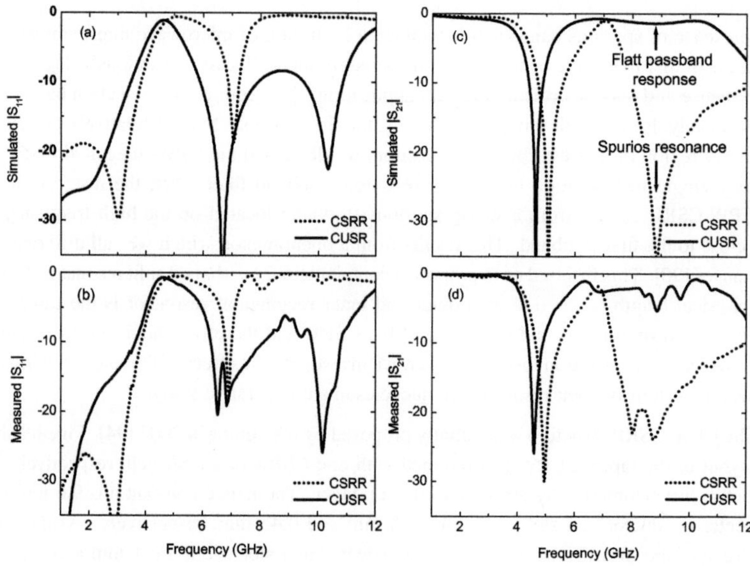

Figure 4.10: Simulated (a) and measured (b) return loss for CPW-CSRR and CPW-CUSR structures. Simulated (c) and measured (d) insertion loss for CPW-CSRR and CPW-CUSR structures.

of 35 μm). The S-parameters of the fabricated structures were measured within the frequency range of 2 to 12 GHz using an HP E8361A vector network analyzer (VNA) with a microstrip test fixture (Wiltron 3680). A thru-short-line (TRL) calibration was performed for the CPW. The return loss is better than 31 dB for the band of interest. Figs. 4.9(a) and (b) depict the dimensions of both the CSRR and CUSR resonators. They have been chosen for operation in the C-band. They have an external radius of $r = 3.6$ mm, a separation of $c = 0.4$ mm, a width of $d = 0.4$ mm, and a metallic strip $g = 0.4$ mm in the case of the CSRR. In the CUSR structure the u-shaped resonator has the same physical length as the "inverted u". The fabricated CPW/CSRR and CPW/CUSR structures are shown in Figs. 4.9(c) and (d), respectively.

To simulate the frequency response through the structures we use commercial software (Ansoft HFSS [74]). The simulated and measured magnitudes of the return and insertion losses for both CSRR and CUSR structures are shown in Fig. 4.10. The dashed curves depict the responses for the CSRR while the solid curves demonstrate the responses for CUSR. For both structures there is a strong main stopband where the real part of the electric permittivity is expected to be negative. The CSRR has a higher level of return loss in the upper band of the stop band filter. Yet, the case of CUSRs the return loss level is improved by 8 dB for frequencies between 7 GHz and 11 GHz. The simulated and measured insertion loss is shown in Figs. 4.10(a) and (c), respectively. It reveals the potential of the CUSR resonator. The simulated and (measured) stopbands are centered at 4.6 GHz (4.6 GHz) and 5.1 GHz (5 GHz) for CUSR and CSRR, respectively. Although, both structures have a very good rejection level and sharp transition edges, there is a strong spurious extended stopband centered around 8.8 GHz, i.e. very close to the main stopband. In the case of CUSRs the first spurious stopband is centered only at 16.3 GHz.

Note that the resonance frequency of the CUSRs is 10% lower than that of the CSRRs. Hence, a miniaturization can be achieved with CUSRs in comparison with CSRRs. The measured data for the return and insertion losses are in good agreement with the simulated results. The small mismatch between the simulated and measured data is attributed to inhomogeneities in the fabricated structures. In summary, the new design shows a much better performance compared with CSRRs of the same physical size.

We further investigated the effect of different geometrical dimensions on the resonance frequency of the structures. Fig. 4.11 shows results obtained from a large set of simulations for CSRR (circles) and CUSR structures (rectangles). It is found that the resonance frequency increases with the separation c and with the width d. The opposite (complementary) effect has been observed for the conventional SRR [89]. As can be seen from Fig. 4.11 the resonance frequency is more sensitive to the width d than to the separation c. For all values of d and c the resonance frequency is smaller for the CUSR structure. This means that the CUSR structure offers a higher degree of miniaturization.

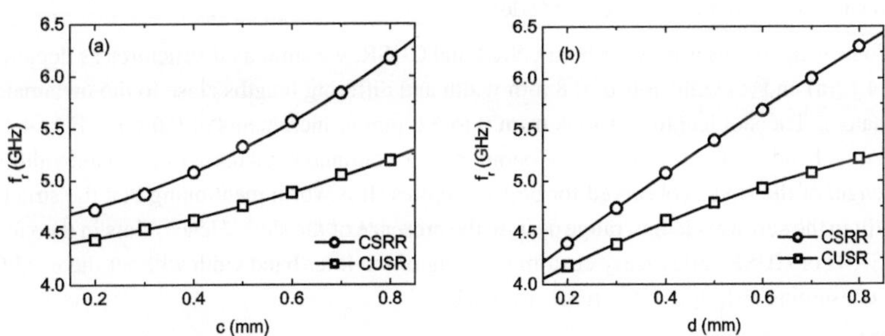

Figure 4.11: Resonance frequency (f_r) vs. separation c (a) and width d (b) of the CSRR and CUSR resonators.

4.2.4 Bandwidth Modifying Slots

It is well known that the bandwidth provided by an SRR structure is rather small and not sufficient for some applications (see e.g. [95]). To widen the overall stop band it has been proposed to use several SRR cells in series, each of which has a slightly different resonance frequency. Going from one to five SRR cells, for example, increases the bandwidth by 30% [48]. Naturally, the CSRR structure has the same limitation because of the above mentioned concept of

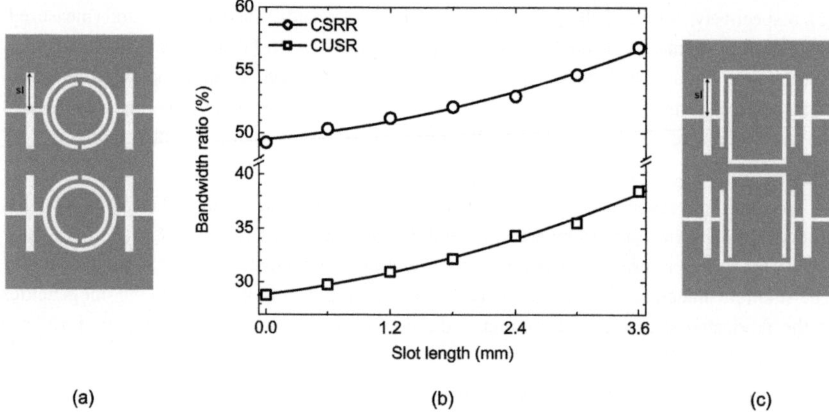

Figure 4.12: (a) CPW-CSRR with nearby slots (b) bandwidth of the stop band filter vs. the slot length for CPW-CSRR and CPW-CSCR structures (c) CPW-CUSR with nearby slots.

duality. We demonstrate here, that the introduction of slots in close proximity of the CSRRs can enhance the bandwidth considerably. These slots can easily increase the width of the stopband by 60% or even more. Hence, the overall structure (a single CSRR plus slots) is still rather small since no further cells have to be added.

To investigate this aspect for both CSRR and CUSR, we simulated structures as depicted in Fig. 4.12(a) and (c) with slots of 0.8 mm width and differing lengths close to the metamaterial resonators. The slot length sl varies from 0 to 3.6 mm in increments of 0.6 mm. Fig. 4.12(b) shows the bandwidth ratio over the sl parameter. A continuous increase of the bandwidth with the length of the slots is observed for both structures. It is worth mentioning that the structures are still in the sub-wavelength range despite the presence of the slots. Hence, slots in the vicinity of CSRRs or CUSRs allow easy custom tailoring of the filter bandwidth without the need for a time consuming redesign of the resonator itself.

The slots effectively act as series reactance and by this extent the rejected bandwidth of the CSRR which has predominantly a capacitive behavior in the region of the resonance frequency. The dominant capacitive behavior of the CSRR resonator is caused by the effective capacitance of the internal and external rings (in this complementary structure the rings consist of air). The underlying physical mechanism can be intuitively understood in the framework of a circuit model. Such as model for the upper half of the structure (note the structure is symmetric) is shown in Fig. 4.13. Although a complete model of the CSRR would be complicated, it is modeled here for simplicity as a resonator containing a capacitance C_r and inductance L_r in parallel (see solid box in Fig. 4.13). In front of the upper CSRR (left of it) we have two slots. One of them is in the center conductor and the other one is in the ground. The slot in the center conductor can be modeled as a series reactance (inductance with capacitance in parallel). The slot in the ground conductor has the same effect, yet, the absolute value of the inductance and capacitance may be a bit different. Together the two slots form a simple

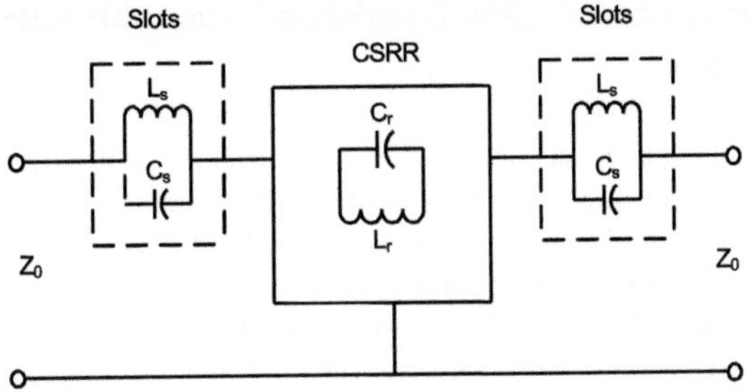

Figure 4.13: Equivalent-circuit model for the CSRR-slots structure.

resonance circuit enclosed in Fig. 4.13 by a dashed box. After the CSRR we have a second pair of slots (right dashed box). According to this model, an electromagnetic wave propagating through the structure will face more impedance for frequencies in vicinity of the resonance frequency. Hence, the transmission is somewhat reduced which in effect increases the rejected bandwidth. The simulated and measured curves for the CSRR structures with $(sl = 3.6\ mm)$ and without slots are shown in Fig. 4.14. It is interesting to note, that the presence of the slots close to the CSRRs leads to a broadening of the stopband which ranges now from 4.1 to 5.6 and, hence, is 60% broader. Note, that also the transmission at the resonance frequency is reduced by 8 dB.

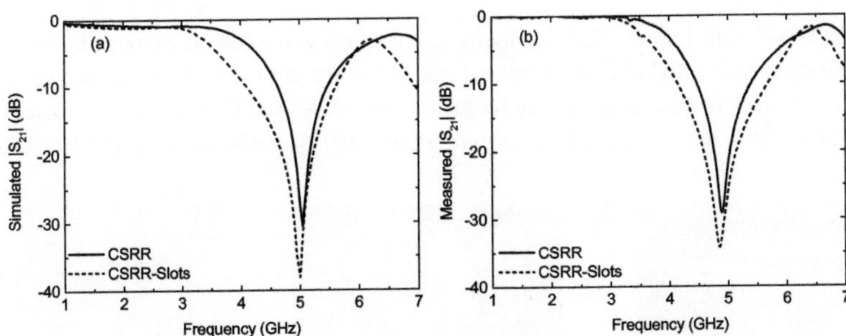

Figure 4.14: (a) Simulated insertion and return losses for a CPW incorporating a single pair of CSRRs (dashed line) and a single pair of CSRRs with slots (solid line) (b) measured insertion and return losses for a CPW incorporating a single pair of CSRRs (dashed line) and a single pair of CSRRs with slots (solid line).

4.2.5 Miniaturization of CPW Bandstop Filter with Spiral Resonators

Many applications throughout microwave technology demand for a high degree of miniaturization. In order to meet this challenge, the concept of spiral resonators will be studied in this section in conjunction with the previously discussed CPW integrated CSRRs and demonstrate the high miniaturization potential which arises from this combination. The spiral resonator was proposed to further reduce the size of the SRRs [96]. We utilize the complementary split rectangle resonators (CSCR) and the complementary spiral resonator (CSR) with multi turns to obtain single layer bandstop filters with a high degree of miniaturization. These structures could be of interest for applications in bandstop filters where miniaturization and compatibility with planar technology are key issues.

Fig. 4.15 shows the layout of the tapered CPW incorporating a unit cell of CSRRs, CSCRs, and CSRs, respectively. The resonators are symmetrically etched into the top layer. The lattice constant a is 8 mm. The FR-4 substrate is employed for all the structures (dielectric constant ϵ_r = 4, loss tangent tan δ = 0.02, and thickness h = 0.5 mm). The dimensions of both the CSRR and CSCR resonators are with an external radius of r = 3.6 mm, a separation of c = 0.2 mm, a width of d = 0.2 mm, and a metallic strip g = 0.2 mm.

To simulate the frequency response through the structures we use commercial software (Ansoft HFSS [74]). The simulated magnitude of the frequency response for the three structures is shown in Fig. 4.16. For all structures there is a strong main stopband where the real part of the electric permittivity is expected to be negative. It reveals the potential of the CSR resonator. The stopbands are centered at 4.08 GHz, 3.36 GHz, and 1.38 GHz for the CSRR, CSCR, and CSR, respectively. Note that the resonance frequency of the CSRs is one-third that of the CSRRs. Hence, a higher degree of miniaturization can be achieved with complementary spiral resonators in comparison with CSRRs. The measurements are shown in Fig. 4.16. They are in very good agreement with the simulated results.

Furthermore, the impact of adding more internal turns to the CSR is examined. Two, four, six, and eight turns of CSRs are compared (the structures are shown as insets in Fig. 4.17). Fig. 4.17(a) depicts the first resonance frequency (left scale) and the corresponding electrical size of the CSR in terms of guided wavelength (right scale). There is saturation in the resonance

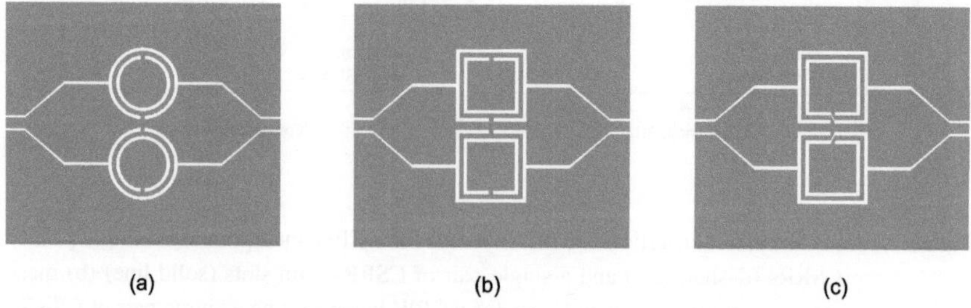

(a)   (b)   (c)

Figure 4.15: Layout of the CPW/CSRR (a), CPW/CSCR (b) and CPW/CSR structures (c).

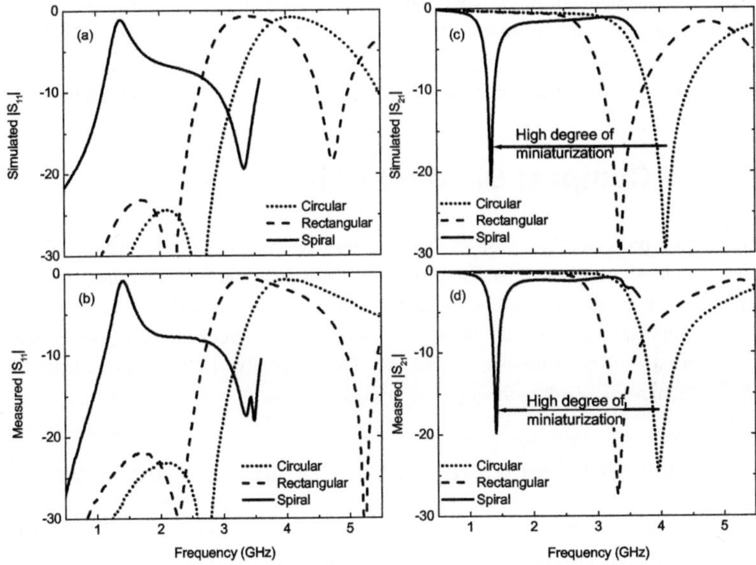

Figure 4.16: Simulated (a) and measured (b) transmission responses for the CPW/CSRR (circular), CPW/CSCR (rectangular), and CPW/CSR (spiral) structures.

frequency with increasing the number of turns. This effect is expected because the resonator inductance and capacitance are saturated. The same behavior has been noticed in [92]. However, a very small electrical size of $\lambda_g/50$ is observed with eight turns CSR. Moreover, it is possible to increase the number of turns by making the metal lines and the spacing between them thinner.

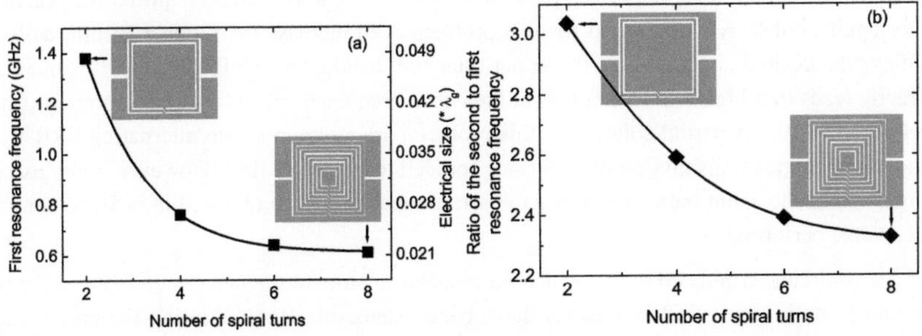

Figure 4.17: (a) Resonance frequency versus number of spiral turns (b) ratio of the second to first resonance frequencies.

All structures show a second resonance (not shown in Fig. 4.16). Fig. 4.17(b) shows the ratio between the frequencies of the second and the first resonance. The ratio decreases with the number of turns. However, it saturates and is above 2.2 even for an eight turn CSR. Thus the second resonance is still sufficiently separated to give a good filter performance. CSR design shows a much better performance compared with CSRRs of the same physical size.

4.3 CPW Compact Bandpass Filters

So far, we have presented how CSRRs can be employed to miniaturize conventional bandstop filters tremendously. Although bandstop filters are very important as mentioned earlier, many modern microwave applications rely on bandpass filters (BPFs). In order to succeed in the field of wireless applications, miniaturization of these versatile devices is mandatory to achieve a high integration density in the overall system. This section will review the recent advances which were achieved regarding this topic in this dissertation.

4.3.1 Narrow BPF Incorporating Bandwidth Modifying Slots

In this section we demonstrate metamaterial-based coplanar waveguide bandpass filters with slots inserted in direct proximity to split ring resonators. The slots enable a narrow bandwidth by varying their lengths. This concept could also be applied to other resonator designs. The proposed structures feature a narrow bandwidth, high stopband suppression and a small electrical size. A dispersion analysis, together with the simulated and measured complex transmission response is employed to evaluate the devices.

Left-handed materials in CPW technology can be obtained by split ring resonators (SRRs), placed on the backside of a substrate, in combination with periodically aligned metallic strips on the upside, connecting the central conductor to the ground planes [48, 97, 98]. The SRRs give rise to the negative effective permeability near their resonance frequency whereas the periodical short circuit strips introduce the negative effective permittivity. The frequency response of such a structure exhibits low insertion losses and a sharp cutoff at the lower band edge. However, the attenuation level above the pass band is usually not as high as required for practical filter applications. A workaround for this problem is to increase the number of unit cells to achieve the desired attenuation at the upper transition band edge. Unfortunately, this increase directly leads to a higher insertion loss at the center frequency as well, so that a trade-off has to be found [90]. A partial solution to this problem, are structures with alternating SRR-wire stages and gapped transmission line sections coupled to the SRRs [99]. However, complex optimization of the shunt inductances and the series capacitance are required in order to achieve acceptable performance.

Short-circuit terminated stubs, which correspond to slots in the center conductor or in the ground plane in case of CPW designs, have been extensively studied during the past decade [100, 101]. In Sec. 4.2.4, this concept was applied to alter the impedance of a negative permittivity only metamaterial by inserting slots close to a pair of complementary SRRs. On this basis, stopband filters with increased rejection bandwidth were demonstrated. In the following,

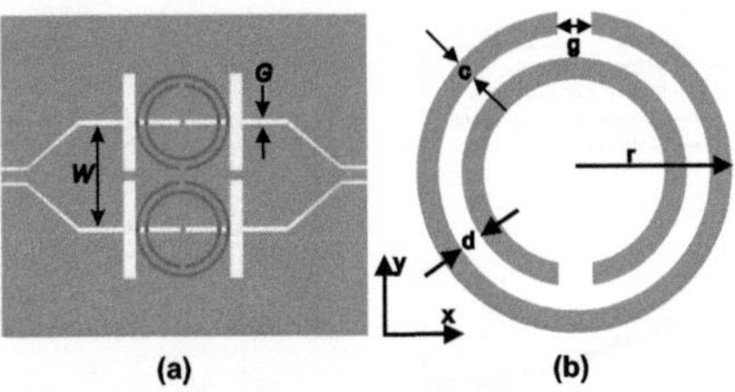

Figure 4.18: (a) Layout of the CPW structure with SRRs on the backside and the slots on the top (b) detailed schematic of a single split ring resonator.

we study the effect of introducing slots in close proximity to the SRRs. Inserting the slots results in a drastically improved attenuation level in the upper stop band. Furthermore, a full control of the bandwidth can be obtained by varying the length of the slots, which enables very flexible filter designs.

The layout of the unit cell of the proposed structure is shown in Fig. 4.18(a). The SRRs are located on the backside of the substrate. On the upside of the substrate, the center conductor and the ground plane are short circuited directly above the center of the SRRs. Slots were etched into the ground plane and the center conductor. Conventional mask/photoetching technique was used to fabricate the structures on a Rogers RO4003 substrate (dielectric constant ϵ_r = 3.38, thickness $h = 0.5$ mm and a copper metallization of $t = 17$ μm). The dimensions of the SRR, as depicted in Fig. 4.18(b), have been chosen so that an S-band (2.6-4 GHz) filter results. The outer radius of the SRRs is $r = 3.6$ mm, the width $c = 0.29$ mm, the separation $d = 0.44$ mm and the gap $g = 0.43$ mm. For the initial experiments a slot width of 0.96 mm and a slot length (sl) of 3.6 mm were employed. Fig. 4.18(a) shows the case with slots both in the ground plane and the center conductor. The lattice constants are $a_y = 8$ mm and $a_x = 9.92$ mm. Finally, the lateral dimensions of the host CPW ($W = 7.61$ mm, $G = 0.39$ mm) have been selected to obtain a 50 Ohm characteristic impedance with a relatively wide center conductor in order to prevent overlapping of the SRR pairs. Tapers have been added to both sides of the CPW to allow the use of a fixture (Wiltron 3680), which avoids measurement errors due to soldered connectors. The measurements have been conducted using a vector network analyzer (VNA) from Agilent (E8361A). Again a CPW thru-short-line calibration kit was employed to calibrate the VNA. It is found that the return loss is better than 40 dB and the insertion loss is bouncing between 0 and 0.02 dB for the thru calibration component.

For frequencies above the resonance region, the SRRs exhibit a dominant capacitive behavior, which results from the effective capacitance between the internal and the external rings. The slot length is kept in the order of a tenth of the resonance wavelength to preserve the size requirements of metamaterials. The slots in both cases can be modeled as a series inductance with lumped resistance. However, it has been shown, that the maximum inductance of the slot in the ground plane is half of the one in the center conductor [100, 101]. Thus, we shall expect less bandwidth for slots in the center conductor than for slots in the ground plane. A detailed equivalent circuit model for SRR loaded left-handed transmission lines is given in [98].

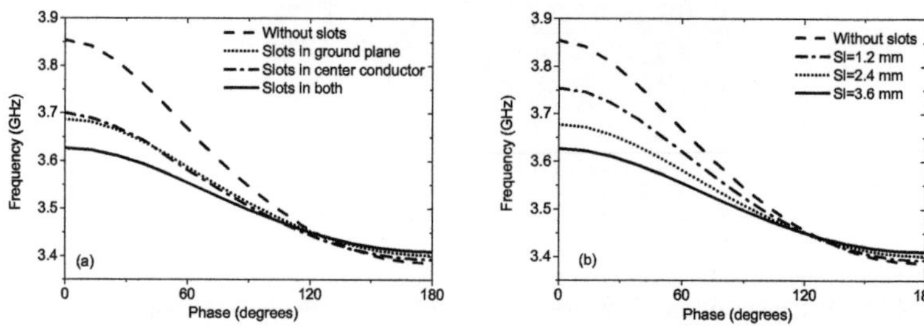

Figure 4.19: (a) Dispersion analysis for one unit cell with different combinations of 3.6 mm slots (b) dispersion analysis for one unit cell with different slot lengths.

In order to predict the qualitative behavior of the device characteristics, the dispersion of the structure is studied on the basis of an Eigenmode analysis employing periodic boundary conditions at the edges of the unit cell. Fig. 4.19(a) shows the results of these calculations for an infinite array of unit cells with different combinations of slots. It shows modes of SRRs without slots (dashed line), with slots in the ground plane (dotted line), with slots in the center conductor (dashed dotted line) as well as with slots in both (solid line). The dispersion graph reveals that the structure has a LHM behavior in the band between 3.38 GHz and 3.85 GHz. Thus, a pass band behavior in the transmission amplitude is expected for the band in which the phase response increases with frequency, indicating a negative phase velocity. Adding slots shrinks the bandwidth successively. While the lower frequency edge of the passband remains nearly unchanged, the higher frequency edge is drastically shifted towards lower frequencies. As expected from [100] and [101], the decrease in bandwidth is stronger for slots in the center conductor than for slots in the ground plane. Having the slots in both the ground plane and the center conductor results in a very narrow bandwidth, which in case of the slot lengths given above, is only 44 % of the bandwidth without the slots. In terms of bandwidth percentage, this value corresponds to a decrease from 13% to 6%. If even smaller bandwidth percentage is required, the slot lengths can be further increased. Moreover, Fig. 4.19(b) depicts the dispersion analysis for a sweep over different slot lengths with the slots being located in both the center conductor and the ground. Increasing the slot length directly reduces the bandwidth as expected from theory.

All following simulations in this section were performed using a commercial 3D full wave electromagnetic simulator (Ansoft HFSS) [74], which is based on the numerical finite element method. Four cases are investigated for the single unit cell. Fig. 4.20 shows the simulated results for SRRs only (dashed line), for SRRs and slots in the ground plane (dotted line), for SRRs and slots in the center conductor (dashed dotted line) and finally for SRRs and slots in both the ground plane and the center conductor (solid line). Fig. 4.20(a) illustrates the transmission amplitude. The 3-dB bandwidth is reduced from 320 MHz in the case of SRRs only to 250 MHz, 190 MHz and 125 MHz, respectively, if slots are present. The phase response is shown in Fig. 4.20(b). Both, the phase and the amplitude response qualitatively agree well with the results

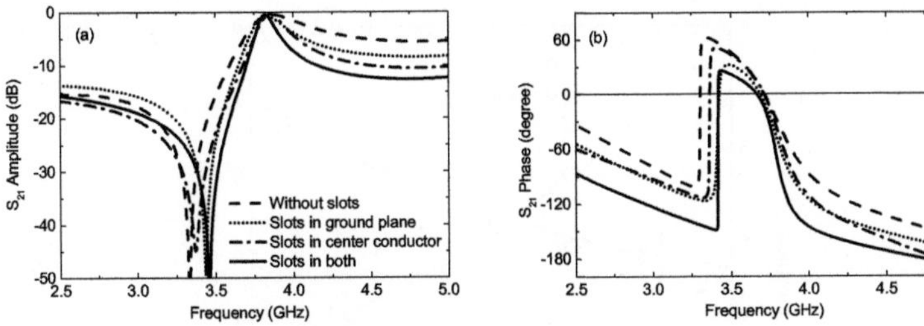

Figure 4.20: Simulated transmission (a) magnitudes and (b) phases for one unit cell with different combinations of slots.

from the dispersion analysis discussed above. Note again, that the simulations of Fig. 4.19 are carried out for an infinite number of cells while in Fig. 4.20 are performed for a single cell. To study the transmission behavior of cascaded unit cells with differing slot lengths, three pairs of SRRs and three slot configurations have been chosen: once without any slots, once with slots of 1 mm length, and finally with slots of 3.6 mm length. The simulated and measured results are shown in Fig. 4.21(a) and Fig. 4.21(b), respectively.

In case of the 3.6 mm slots, a very narrow bandwidth of 125 MHz at a center frequency of 3.8 GHz could be achieved, which is 30% of the bandwidth without slots. This value corresponds to a bandwidth percentage of less than 3.3%. The suppression level at higher frequencies above the passband is almost doubled compared to the case without slots. The measurements confirm the simulated results. Employing more expensive microwave substrates can further reduce the insertion loss, if this parameter should prove critical to the design. Simulations for Rogers RO3003 as substrate have shown a decrease in the insertion loss of almost 1 dB, which corresponds to 28%. It is worth noting that the actual size of the entire structure is only 0.48 λ_g by 0.25 λ_g, where λ_g is the guided design wavelength and the structures still hold a high minia-

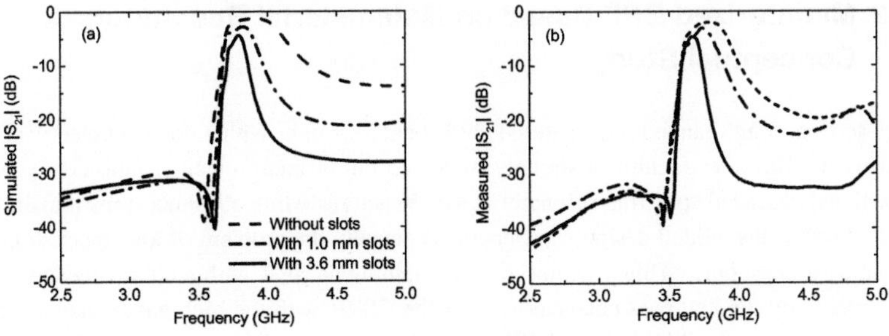

Figure 4.21: (a) Simulated transmission magnitude (S_{21}) (b) measured transmission magnitude (S_{21}) for three unit cells with different length of the slots.

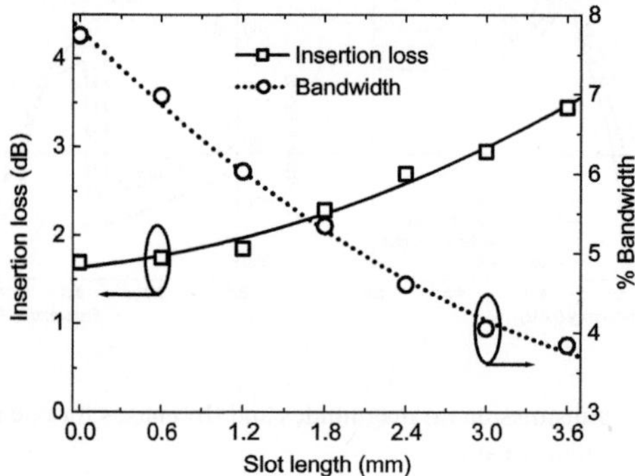

Figure 4.22: Simulated insertion loss (rectangles, left scale) and simulated bandwidth percentage (circles, right scale) over the slot length for a three cell bandpass filter. The dashed and solid lines show the corresponding second order polynomial fits.

turization potential as discussed in [102]. Furthermore, the bandwidth of the proposed CPW bandpass filter can be custom-tailored to the device requirements by varying the slot length, enabling highly flexible filter designs. This design flexibility is a strong advantage over the devices demonstrated in [99] and [103], where a complete filter redesign is necessary to alter the bandwidth. Fig. 4.21 clearly suggests, that a tradeoff between the bandwidth and the insertion loss exists. This aspect is illuminated a little further in Fig. 4.22 by depicting the simulated insertion loss (rectangles, left scale) and the bandwidth percentage (circles, right scale) over the slot length. The solid and the dashed line correspond to the second order polynomial fits. The insertion loss increases and the bandwidth decreases towards higher slot lengths. For example, halving the bandwidth, leads to an added insertion loss of almost 1.8 dB.

4.3.2 Miniaturized BPF Based on Metamaterial Resonators-A Conceptual Study

In this section, a left-handed metamaterial CPW bandpass filter with a compact electrical size is proposed. The filter features a strongly improved out-of-band rejection ratio compared to conventional cascaded split ring resonator (SRR) structures while all dimensions remain well below a third of the guided design wavelength. Especially the problem of low upper stopband attenuation is overcome. This is achieved by combining an SRR with a complementary SRR cell. Combining the low pass characteristics of the CSRR with the high pass behavior of the SRRs allows a very flexible design of BPFs, with the resonance frequency of the SRR defining the lower band edge and the resonance frequency of the CSRR defining the upper band edge. Moreover, this approach enables flexible filter designs for a broad range of bandwidth ratios with small insertion losses at the center frequency of the device.

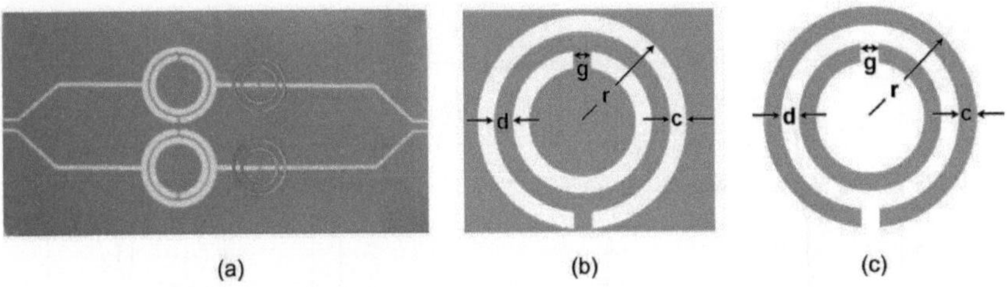

Figure 4.23: (a) SRR/SL-CSRR loaded the CPW fabricated structure. Schematic of CSRR (b) and SRR (c) with relevant dimensions.

The sample geometry is depicted in Fig. 4.23. The CPW tapers are employed to exclude measurement errors due to soldering connectors to the devices. The taper is verified through simulation and measurement to provide maximum matching between the two sides of the CPW. The unit cell of the structure consists of a pair of SRRs with shunt strip lines (SRR/SLs) and a pair of CSRRs. The dimensions of the SRRs are chosen such that the device operates in the C-Band around 4 GHz. Both SRRs and CSRRs have the same width of $c = 0.42$ mm, a separation of $d = 0.38$ mm and a gap of $g = 0.36$ mm. The outer radius r is 3.2 mm and 3.6 mm for the SRRs and CSRRs, respectively. A standard mask/photoetching technique is employed to fabricate the samples using a commercial low cost FR-4 substrate (dielectric constant $\epsilon_r = 4$; loss tangent tan $\delta = 0.02$, thickness $h = 0.5$ mm).

Simulations and measurements are performed for three selected structures. They comprise a single stage of each, SRR/SL and CSRR as well as a combination of both (SRR/SL-CSRR). A commercially available 3D full-wave solver (Ansoft HFSS [74]) based on the finite element method with adaptive meshing is employed to calculate the S_{21}-parameter, which corresponds to the transmission insertion loss of the structures. Accompanying measurements were performed using a vector network analyzer (HP E8361A) with a microstrip test fixture (Wiltron 3680). The vector network analyzer was calibrated using a thru-short-line calibration kit. The return

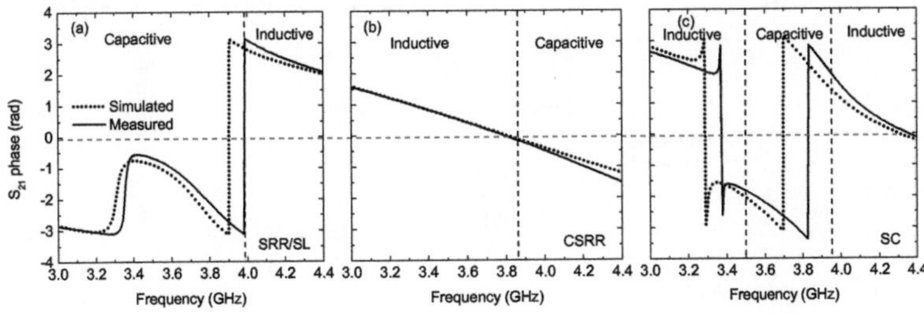

Figure 4.24: Simulated and measured transmission phase for SRR/SL (a), CSRR (b) and both structures (c).

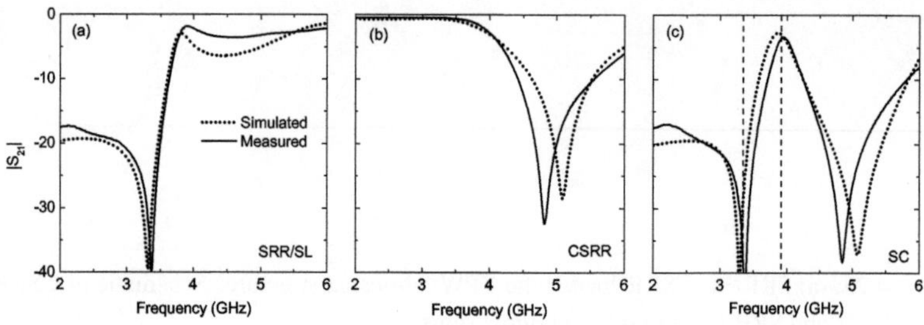

Figure 4.25: Simulated and measured return loss for the SRR/SL (a), CSRR (b) and both structures (c).

loss is better than 26 dB for the band of interest with an insertion loss variation of 0.02 dB. The simulated and measured phase, return loss, and the insertion loss are given in Fig. 4.24-Fig. 4.26, respectively.

The simulations qualitatively agree very well with the measurements. Vertical dashed lines, connecting the simulation and measurement graphs at selected frequencies reveal only small quantitative discrepancies, which can be ascribed to manufacturing inaccuracies connected to the wet etching process. Fig. 4.24 and Fig. 4.25 allow an intuitive understanding of the overall BPF concept presented here. The low pass magnitude response of the CSRRs together with the high pass characteristics of the SRR/SLs results in a bandpass response of the combined SRR/SL-CSRR BPF with sharp cut-off and high stopband attenuation as depicted in Fig. 4.25. Investigating the phase responses of the single elements and the overall structure reveals the optimization options in terms of impedance matching and insertion loss. The phase response

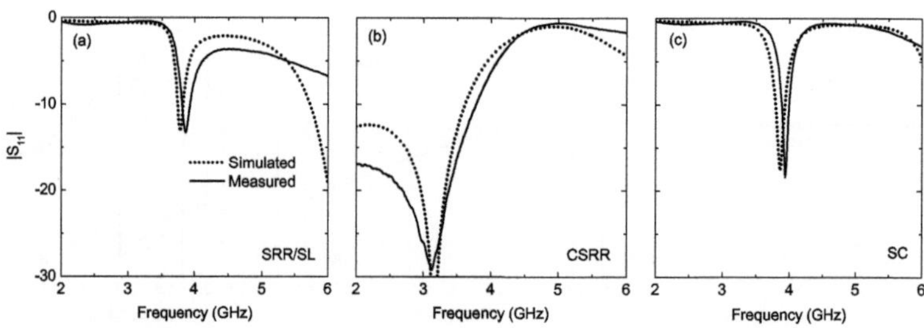

Figure 4.26: Simulated and measured insertion loss for the SRR/SL (a), CSRR (b) and both structures (c).

of the SRR/SLs is dominantly capacitive for the lower stopband. At the resonance frequency, a phase flip occurs, after which the SRR/SLs behave dominantly inductive. The CSRRs exhibits a gradually decreasing phase response over the whole relevant frequency range. Close to the center frequency, the behavior of the CSRRs changes from inductive to capacitive which helps to compensate for the high inductive reactance of the SRR/SLs after their phase flip resulting in lower insertion loss in the passband.

The simulated and measured insertion loss at the center frequency of $f_c = 4$ GHz is 2.7 dB and 3.4 dB, respectively. The out-of-band rejection is higher than 19 dB in the simulation and 17 dB in the measurement. Keeping in mind, that the whole structure has a geometrical outline of only 0.29 λ_g by 0.29 λ_g (with λ_g as guided wavelength) and is based on standard FR-4 substrate material, the performance is superior to previously demonstrated devices which employ cascaded SRR cells. The simulated and measured return loss for all three structures is shown in Fig. 4.26. At the center frequency, the measured and simulated value for the combined SRR/SL-CSRR is 18.9 and 17.6 dB, respectively. Thus, only 1.5 % of the forward power is reflected back to the source, which further confirms the high potential of this filter concept.

Simulations with a more expensive microwave substrate (Rogers RO3003, dielectric constant $\epsilon_r = 3$, tan $\delta = 0.0013$, thickness $h = 0.127$ mm) have also been performed. In this case, a 60% (1.6 dB) reduced insertion loss and a 4 dB enhanced out-of-band rejection result. In order to obtain the same center frequency of the FR-4 sample, the radius of the split rings was reduced to 2.8 mm to compensate for the difference in substrate thickness and the dielectric constant. Experimentally, a reduction in the insertion loss at the center frequency of 55 % (1.5 dB) is found, which is close to the calculated value. This provides an alternative to the FR-4, if the increased performance is required. However, it should be noted that the presented design allows very narrow passbands usually only available for filters constructed in more expensive technologies, e.g. based on high-temperature superconductor thin films.

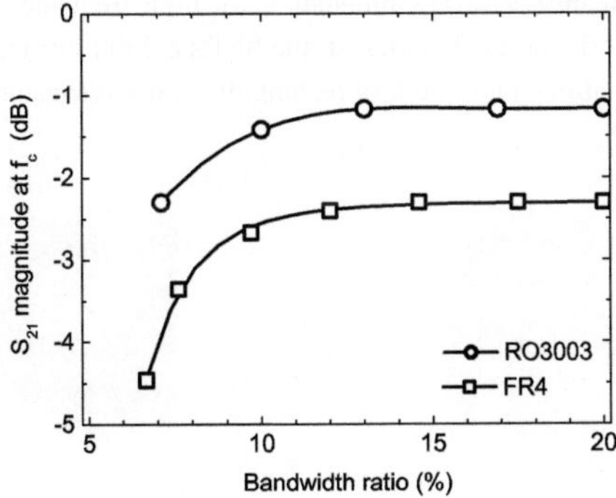

Figure 4.27: Simulated S_{21} magnitude at the center frequency versus the bandwidth ratio for a sweep of the outer radius of the SRR.

Fig. 4.27 analyzes the dependence of the insertion loss on the bandwidth ratio (for FR-4 and RO3003 substrates). To obtain corresponding pairs of these two parameters, the outer radius of the SRRs was swept from 3.0 mm to 3.6 mm and 2.6 mm to 3.2 mm in steps of 0.1 mm for the FR-4 and the RO3003, respectively. As stated in [61], the resonance frequency of the SRRs is proportional to the inverse square root of the outer radius. Thus, an increase in the radius of the SRRs lowers their resonance frequency, while the resonance frequency of the CSRRs remains unchanged. This leads to a widening of the bandwidth of the overall structure. The bandwidth ratio is defined as the relation of the bandwidth over the center frequency so that the bandwidth ratio is increased as well.

For high bandwidth ratios, the losses at the design frequency remain low. This can be understood, if the overall bandpass design is again considered as the superposition of a low pass (CSRRs) and a high pass (SRRs) filter. As long as the design frequency of the bandpass filter lies within the passband of both, the high- and the low pass, the sensitivity of the insertion loss on the bandwidth ratio remains very small. This changes when the design frequency of the bandpass filters lies in the transition region of passband and stopband of the low pass and the high pass filter, respectively. Reducing the bandwidth ratio a little further would directly result in drastically increased losses. This distinct behavior is depicted in Fig. 4.27. In the case of the presented design, the losses remain low for bandwidth ratios of down to 10%. This enables custom tailored filter designs by varying only a single geometrical parameter of the structure, namely the outer radius of the SRRs. If an application should require a bandwidth ratio below 10% and the insertion loss would be too high, additional SRR or CSRR cells could be cascaded to provide better performance.

4.3.3 Very Compact BPF Based on Spiral Metamaterial Resonators

In the last section, we have demonstrated a BPF based on metamaterial resonators with an electrical size of one-third of the guided wavelength. It is based on combining conventional SRRs with CSRRs. CSRRs feature a strong attenuation for high frequencies while low frequency components pass with little losses. In contrast, the SRRs exhibit just the inverse behavior. The aim of this section is to utilize multiple turn rectangular spiral resonators (RSRs) and the com-

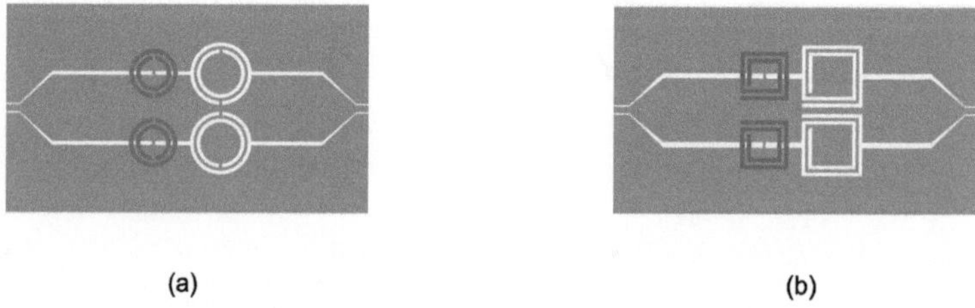

(a)   (b)

Figure 4.28: Layout of the structure (SRR/SL-CSRR) CPW (a) and the proposed structure (RSR/SL-CRSR) CPW (b), respectively.

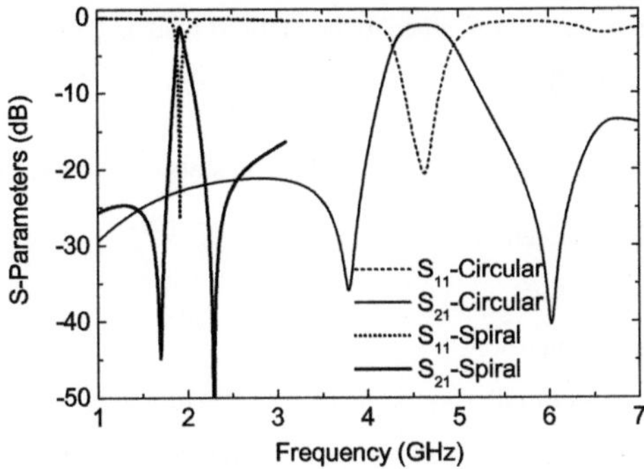

Figure 4.29: S-parameters for the SRR/SL-CSRR and RSR/SL-CRSR.

plementary structures (CRSRs) to achieve a drastically improved miniaturization [104]. Simulations are performed for two designs shown in Fig. 4.28: an SRR/SL in series with a CSRR and an RSR/SL in series with a CRSR. An RO3003 substrate with a 0.127 mm thickness was used for all simulations. The simulated return (dashed) and insertion (solid) loss for both structures is depicted in Fig. 4.29. While the center frequency of the circular structure is 4.6 GHz with an insertion loss of 1 dB, the rectangular structures exhibit a resonance at only 1.92 GHz with a comparable insertion loss of 1.3 dB. The corresponding 3-dB bandwidth ratio is 15 % and 5 % for the circular and rectangular spiral structure, respectively. Although the electrical size of the circular structures is relatively small with only 0.29 λ_g by 0.29 λ_g (with λ_g as guided wavelength), the rectangular spiral structure are even further miniaturized with electrical outline dimensions of only 0.12 λ_g by 0.12 λ_g. Thus, less than a fifth of the area is required, while the bandpass performance is fully maintained.

In summary, a compact coplanar left-handed metamaterial bandpass filter is demonstrated, which comprises an RSR/SL and a CRSR cell. The combination of these two elements yields a resonator design with extremely small electrical outline dimensions and a low insertion loss. Compared to conventional SRR filters, a miniaturization of a factor 5.84 with regards to the required board area was achieved. Moreover, the bandwidth ratio is just one-third despite the fact, that the insertion loss increased by merely 0.3 dB. Due to their superior properties, bandpass filters based on this concept show a great potential for many applications, e.g. in transceiver front-end designs of high-speed communication systems.

4.4 Summary

In conclusion, we have discussed recent advances in the field of planar metamaterials. Especially, recently proposed planar metamaterial resonators for high performance, small footprint filters integrated in coplanar waveguide technology were considered. The following list, illuminating briefly the characteristic strengths and weaknesses of the previously discussed concepts, shall serve both as summary and reference to the inclined reader.

Part 1: CPW compact BSFs

- Complementary split ring resonator (CSRR) with slots (Section 4.2.1, see also [94])

    - High stopband rejection

    - Low passband losses

    - Small outline dimensions

    - Slots allow custom-tailoring of the filter bandwidth

    - Spurious resonance in the higher passband due to difference in inner and outer resonator arm length

- Complementary u-shaped split resonator (CUSR) (Section 4.2.3, see also [105])

    - High stopband rejection

    - Low passband losses

    - Small outline dimensions

    - No spurious resonance in the higher passband due to equal resonator arm lengths

- Complementary spiral resonators (CSRs) (Section 4.2.5, see also [106])

    - High stopband rejection

    - Low passband losses

    - Extremely compact size

    - Well suited for frequency selective surface designs with small resonator spacings

    - Spurious resonance due to differing resonator arm lengths.

    - But: still a good separation between spurious resonance and main resonance is achieved

Part 2: CPW compact BPFs

- Narrow BPF incorporating bandwidth modifying slots (Section 4.3.1)

    - Low passband losses

    - High stopband rejection

    - Rather narrow passband

    - Full control of the bandwidth

    - Small footprint

    - Bandwidth shrinking as required by controlling the slot length

- Combined CSRR/SRR with strip lines in series (Section 4.3.2, see also [107])

    - Low passband losses
    - High stopband rejection
    - Moderate outline dimensions
    - Well suited for frontend designs

- Very compact BPF based on spiral metamaterial resonators (Section 4.3.3, see also [104])

    - Low passband losses
    - High stopband rejection
    - Very compact size

5 Dispersion Analysis of CPW LHM

The previous chapter illustrates many novel planar resonators for microwaves circuits. Although microwave filters are often physically very compact, they can contribute a lot to the delay in microwave circuits. Devices incorporating coplanar waveguide (CPW) left-handed media (LHM) with negative index and negative group delay could in principle be used to compensate for positive group delays caused by other devices in a communication system. Moreover, the study of the group delay is very important in radar and ultra-wideband communication systems. The frequency content of a signal pulse is complex and can span a bandwidth of one GHz or more. When the pulse is processed, its spectrum has to be treated uniformly over the entire frequency band, otherwise distortion will, for example, render radar range measurements inaccurate. Besides, group delay issues are crucial in wireless communication systems, which make extensive use of bandpass filters. The bit rate of such systems is highly affected by the group delay of the system's components, especially for wide bandwidth systems. Nearby symbols can run into each other, and therefore contribute to symbol errors.

Therefore, studying the group delay behavior for microwave filters is quite important from a system perspective. This chapter presents a combined theoretical and experimental study on the group delay behavior of four coplanar waveguide structures loaded with pairs of split-ring resonators (SRRs) and strip lines (SLs). We aim to study the behavior of the group delay in the planar left-handed medium. However, we first concentrate on SRRs and strip lines in free-space with dimensions identical to that of the planar structures. The transmission characteristics of this conventional free-space LHM will serve as a reference for the planar structure studied later in the chapter. While we perform only simulations for the free-space structures, the planar structures are investigated experimentally and numerically. Simulations and measurements agree very well in the magnitude, phase and associated group delay of the transmission response. This study can be taken as a template technique for any other structure of interest.

The following section gives a brief introduction to CPW LHM. The section after is devoted to the physical meaning of phase delay and group delay. Then, section three describes the structural details as well as some simulation details. In section four, a dispersion analysis for both types of structures is performed. Section five describes the simulated results and the experimental measurements. Finally, the chapter ends with a short summary.

5.1 Coplanar Waveguide Left-Handed Media

The coplanar waveguide offers several conceptual advantages over the conventional microstrip line as discussed in Sec. 2.4.2. The most important benefit is that the CPW facilitates easy shunt as well as series surface mounting of active and passive devices. Hence, dispersion analysis for such structures is very important from any system perspective which uses these structures.

An important application of an SRR is to serve as load for a CPW as presented in Sec. 2.4.2 [48]. A low loss substrate should be used to get a good bandstop response in transmission. If a strip line shunt interconnection between signal and ground conductors of the CPW is added, bandpass behavior is obtained [97]. This structure is essentially the same as the free-space SRR-wire medium of Smith et al. [58], but it is planar and the fabrication procedure only requires a standard mask/photoetching technique. The structure is very useful for the elimination of frequency parasitics and undesired bands in microwave and millimeter-wave devices. It can be used as a bandstop filter by using the SRRs only, or as a bandpass filter by using SRRs and SLs [97]. Compatibility with monolithic microwave integrated circuit or printed circuit board fabrication technologies as well as compactness (derived from the fact that ring dimensions are electrically small), are other key aspects that are likely to lead to applications of these structures in miniaturized microwave and millimeter wave circuits [97].

5.2 Physical Meaning of Phase Delay and Group Delay

Let us very briefly recall group and phase delay from chapter two. The group delay (τ_g) is defined as the negative rate of change of the insertion phase (ϕ_{21}) with frequency:

$$\tau_g = -\frac{\partial \phi_{21}}{\partial w} \tag{5.1}$$

Group delay is sometimes also called "envelope delay". It is the time delay of the amplitude envelope of a narrow group of frequencies around that specific frequency. In contrast, phase delay is the time delay of a sinusoid at a specific frequency assuming that the sinusoid is constant for all time. Phase delay is related to the phase of the system by the following formula:

$$\tau_p = -\frac{\phi_{21}}{w} \tag{5.2}$$

It is obvious that when the phase (ϕ_{21}) is linear with frequency, both the phase delay and the group delay lead to a constant delay for all frequencies. When the phase is nonlinear with frequency, neither the phase delay nor group delay is constant with frequency [53].

In a network with negative group delay the insertion phase increases with frequency over a certain frequency band. This group delay is the opposite of that obtained from a conventional delay line. Accordingly, the peak of the output waveform arrives before its counterpart at the input [108]. Negative group velocity or superluminal velocity was initially considered as unphysical, and perhaps a mere mathematical artifact [109]. Decades later, in 1982, Chu and Wong [110] were the first to experimentally demonstrate and prove the analytical work of Garret and McCumber from 1970 [111]. They demonstrated the existence of abnormal group velocities for picosecond laser pulses propagating through the excitonic absorption line of a GaP:N sample [10]. Similar effects can be observed in photonic crystal devices [112, 113, 114] as well as in the propagation of electronic signals [115] (A good overview on the history of negative group velocity can be found in [53, 89].)

In 2004, however, Woodley and Mojahedi [89] were the first to study the group delay for the conventional LHM configuration, i.e. split-ring resonators combined with thin strip lines in free space [7] using simulations and experiments for three and four unit cells. Within the region of negative group delay with a bandwidth of 250 MHz a maximum negative group delay of -1.2 ns was observed experimentally [89].

On the other hand, a transmission line loaded with lumped series impedance and shunt inductance was found to exhibit a negative group delay of about -3.1 ns [116]. Today, the term "dispersion engineering" reflects our desire to synthesize and control dispersive effects, and in particular their associated signs, as manifested by the phase and group delay [53].

5.3 LHM Structures

The dimensions of the split-ring resonator (SRR) structures are designed for operation in the S-band between 2.6 and 4 GHz. They have an external radius of $r = 3.6$ mm, a width of $c = 0.27$ mm, a separation of $d = 0.3$ mm, and a gap $g = 0.43$ mm, see Fig. 5.1. Finally, the width of the strip lines SL is 0.43 mm. The dimensions have been selected carefully to have the plasma frequency for the SL greater than the magnetic resonance frequency of the SRR. This should result in a LHM with a pass band behavior starting at the magnetic frequency. The effective permittivity and permeability are negative around this frequency (see Sec. 2.1.5 for more details). FR-4 substrate with a dielectric constant $\epsilon_r = 4$, loss tangent $tan\delta = 0.02$, thickness $h = 0.5$ mm, and copper cladding of 35 μm on both sides is used through the whole simulated and experimental work in this chapter.

The unit cell of the free space structure is shown in Fig. 5.2(a). Perfect magnetic conductor boundary conditions are applied on the x-faces and perfect electric conductor boundary conditions are used on the z-faces of the unit cell to supply natural propagation conditions for a z-polarized plane wave impinging on the structure in y-direction. This wave efficiently excites the negative permittivity behavior. In the direction of propagation (y-axis) the unit cells are

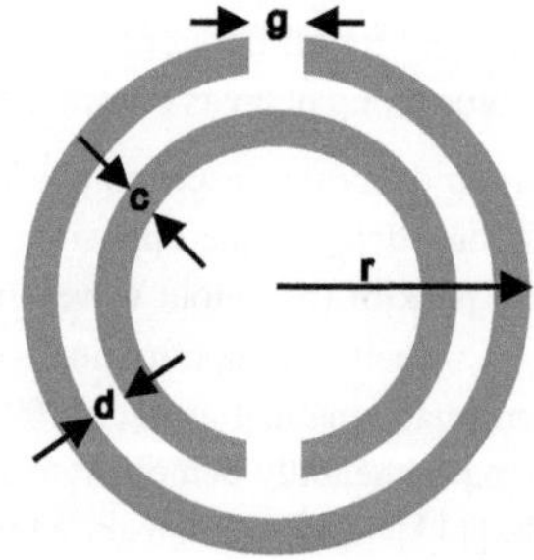

Figure 5.1: Schematic of a split ring resonator with dimensions.

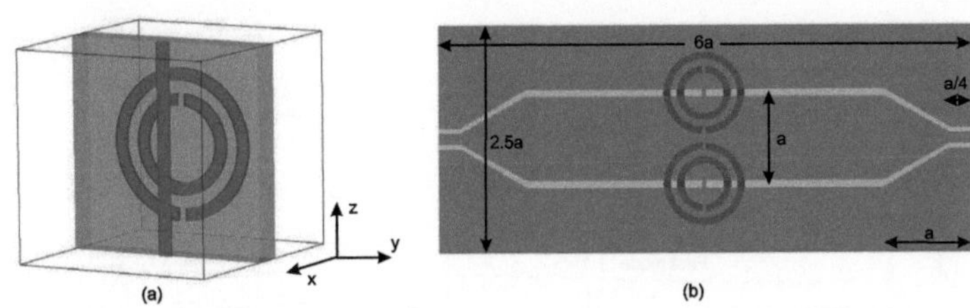

Figure 5.2: (a) SRR unit cell structure in free-space. (b) Layout of the CPW structure.

repeated 2, 3, and 4 times so that transmission through structures comprising 1, 2, 3, and 4 unit cells are simulated.

The CPW LHM structure was first proposed in 2003 [97]. The structure is displayed in Fig. 5.2(b). It consists of a host coplanar waveguide on top of the substrate, which is comprised of a center conductor and two ground planes on either side separated by slots. SRRs are symmetrically placed at the back of the substrate while thin strip lines connect the signal to the ground at the positions which coincide with the centers of the SRRs. The dimensions of the SRRs are the same as that of the free-space structures. In Fig. 5.2(b), the lattice constant a is 8 mm while the center conductor and slot width are 7.61 mm and 0.39 mm, respectively. At the end of the structure the dimensions of the center conductor and slot width are 1 mm and 0.14 mm, respectively. These lateral dimensions have been chosen to obtain a characteristic impedance for the host line of Z_0=50 Ω, and hence to optimize transmission in the pass band. The CPW tapers are added at the edges of the structure to exclude measurement errors due to soldering connectors to the devices. The taper was verified through simulation and experiment to give maximum matching between the two sides of the CPW. An in-house standard mask/photoetching technique is used to fabricate the structures on FR-4 substrate. The structure masks were oversized by 30 μm to compensate for over etching. One should expect some little shifts in the resonance frequency of the rings due to inhomogeneities in the ring dimensions.

In the following two sets of four structures constructed by using one to four SRR-SLs unit cells as explained in Fig. 5.2 will be investigated. For the free space structures only simulations are presented using the boundary conditions as discussed earlier. In the case of the planar CPW structures the simulations are compared to measurements. The dimensions of the SRRs are the same in all eight structures and the group of SRRs is always centered. The center-to-center distance between the SRRs (lattice constant) is 8 mm.

5.4 Dispersion Analysis

In order to predict the qualitative behavior of the dispersion characteristics, we study the the dispersion on the basis of an eigenmode analysis of the structure employing periodic boundary conditions (master-slave boundaries) at the CPW ports. Fig. 5.3 shows the results of the disper-

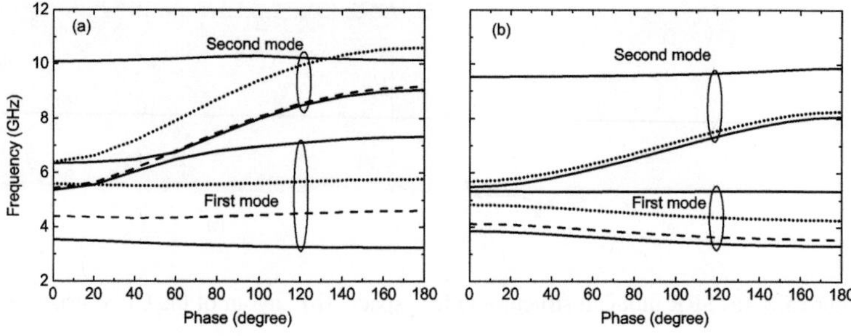

Figure 5.3: Dispersion analysis of the structure (a) in free space and (b) incorporated with the
CPW. Dashed lines show the case of only outer rings, dotted lines show the case of
only inner rings, and the solid lines represents the case of outer and inner rings.

sion analysis for the free space structure in (a) and for the planar structure in (b), respectively.

These figures show modes of SRRs with only the outer rings (dashed lines), only the inner
rings (dotted lines) as well as of SRRs with outer and inner rings (solid lines). In Fig. 5.3(a)
it can be noticed that the structure has a LHM behavior in the band between 3.25 GHz and
3.55 GHz. So, a pass band behavior is expected in the transmission amplitude, for which the
phase response increases with frequency indicating a negative phase velocity. The associated
group delay behavior should reflect this by exhibiting negative group delay starting at the lower
band edge. For the planar structure, the dispersion analysis is shown in Fig. 5.3(b). It shows
almost the same behavior as the structure in free space. The LHM behavior occurs in the range
between 3.32 GHz and 3.84 GHz. The offset in the bandgap is attributed to the proximity
between SRR and the CPW. If the structure contains both outer and inner rings one notices two
additional modes as apparent in Fig. 5.3. This is expected as a result of the coupling of a pair of
resonators, which causes the eigen-frequencies of the coupled system to shift apart. In the case
of the planar structure (Fig. 5.3b), they are narrow band-limited modes at around 5.5 GHz and
9.5 GHz. Although not of particular interest for this work, this explains the occurrence of the
second mode at 5.5 GHz for the planar structures in the transmission results (cf. Figs. 5.5- 5.7).

5.5 Transmission Results

We employ the commercial software (Ansoft HFSS [74]) to simulate the transmission through
the structures (See Sec. 3.1.2). A very high mesh resolution is used by specifying the maximum
change in the S-parameters between two successive iterations to be 0.3%. Moreover, a fine
discrete sweep is performed with a 0.02 GHz step size for the band between 3-4 GHz, where
rapid changes are expected. The S-parameters of the fabricated structures were measured within
the frequency range of 2 to 6 GHz using an HP E8361A vector network analyzer (VNA) with a

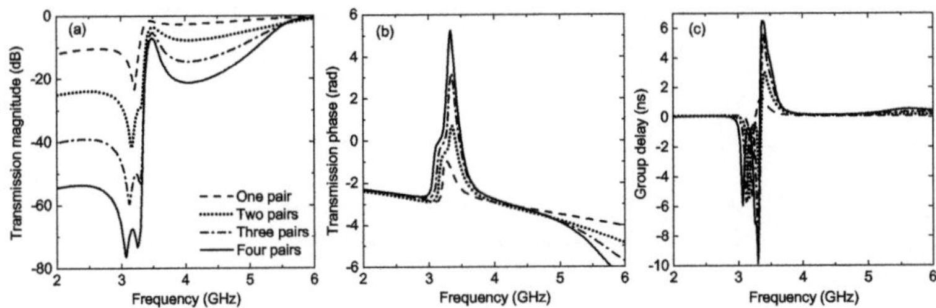

Figure 5.4: Simulated transmission magnitude (a), transmission phase (b), and group delay (c) for one, two, three, and four SRRs-strip lines pairs (free-space structures).

microstrip test fixture (Wiltron 3680). The signal is transmitted through the structure and then to the second arm of the fixture and after that to the VNA again. We put absorbers 3 cm under the SRRs to exclude any interference effects from the fixture itself and the environment. A thru-short-line calibration was performed for the CPW as decribed in Sec. 3.3. The return loss is better than 26 dB for the band of interest (see Fig. 3.10). The phase response for the thru and line calibration sets are checked to verify the quality of the phase measurements.

Fig. 5.4(a) shows simulated magnitude, phase, and group delay for the free space SRR-SL structure shown in Fig. 5.2(a). Clearly, there is one resonance in the single unit cell, while multiple resonances occur for higher number of the cells. For two unit cells, the second resonance starts at a frequency a little bit higher than the first one, while the first resonance shifts to lower frequencies with increasing number of pairs. This behavior can be explained as we have intensified the main magnetic resonance (the first one) by adding more cells with the same dimensions. The phase response in Fig. 5.4(b) shows the LHM behavior increasing in its value inside the resonance band due to the increasing number of cells. The group delay in Fig. 5.4(c) shows one (for a single pair of SRR-SLs) or two distinct negative peaks corresponding to the positive slopes in the phase.

All these results are consistent with the expectations from the dispersion analysis and with the results in the literature using lumped elements [53] and using SRR-SL structures [117]. Fig. 5.5(a) shows the simulated transmission magnitude for the four planar structures. As expected, we observe for all structures a transmission maximum. Around this maximum both the real part of the electric permittivity and the real part of the magnetic permeability are negative. Table I shows all the values of the maximum transmission and the corresponding frequencies. The losses increase with the number of pairs due to the add-on losses originating from the SRRs. Furthermore, an increase in the number of pairs leads to a sharper transmission band. As expected from the dispersion analysis, we note a weak mode at 5.5 GHz, which appears for all structures. It is ascribed to the coupling between the outer and the inner SRRs as stated earlier.

From Fig. 5.5(a) one can notice consistent shapes of the amplitude response curves that have either even or odd numbers of SRR-SL pairs. This behavior is attributed to different field

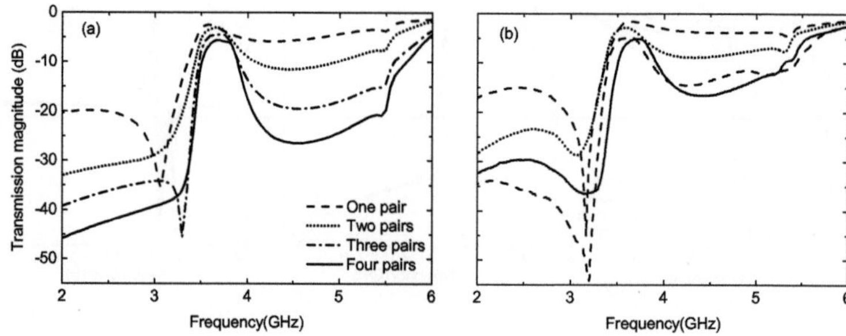

Figure 5.5: Simulated (a) and measured (b) transmission magnitude for one, two, three, and four SRRs-strip lines pairs (planar CPW structures).

distributions in the structures in the vertical and the horizontal direction. The main differences between free-space SRR-SL structures and the CPW-based structures are the size of the short circuit strip and the existence of a current flowing on the CPW, which distorts the magnetic field compared with the homogeneous field in the free-space configuration. Fig. 5.5(b) shows the experimentally obtained transmission magnitude. The data are in excellent agreement with the simulated results. The quantitative agreement is shown in table 5.1. The weak mode at 5.5 GHz also appears in all measurements.

The simulated and measured phase response for the SRR-SL-loaded CPW is given in Fig. 5.6 and the associated group delay is shown in Fig. 5.7. The consistency between simulated and measured phase and associated group delay is very good. This is a confirmation for the simulation strategy and experimental process that was chosen. The phase response for the conventional CPW line is a linear relation, the slope of which represents the group delay. The phase and the associated group delay change as the CPW line is loaded with the pairs of SRRs and SLs. For odd numbers of SRR-SL pairs they show a peak in the phase characteristics giving rise to a negative group delay. The peak is missing for even pair numbers. For odd pair numbers the negative group delay occurs slightly above 3 GHz. The group delay is negative in a rather narrow band. Experimentally we observe bandwidths of 190 MHz and 290 MHz for

	$|S_{21}|$ maximum (dB)		Associated frequency (GHz)	
Number of pairs	Simulated	Measured	Simulated	Measured
One	-2.1	-1.5	3.6	3.62
Two	-3.2	-2.8	3.66	3.6
Three	-4.5	-4.7	3.66	3.64
Four	-5.8	-5.8	3.66	3.65

Table 5.1: Comparison for the transmission peak and the associated frequency from the simulation and measurements.

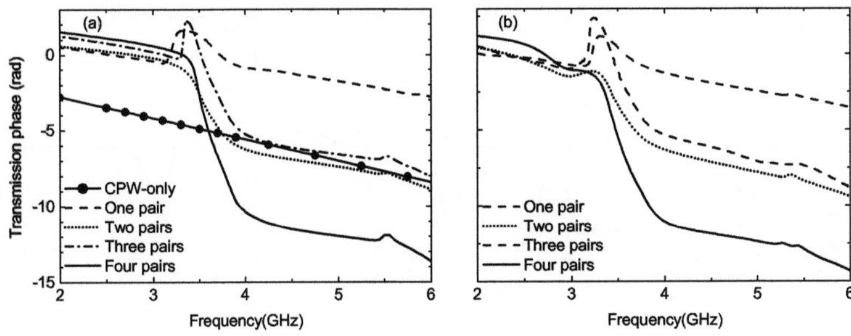

Figure 5.6: Simulated (a) and measured (b) transmission phase for one, two, three, and four SRRs-strip lines pairs (planar CPW structures).

the structures with one and three pairs, respectively, while we obtain 160 MHz and 150 MHz in the simulation. The lowest values of group delay are about -4.7 ns and -8.5 ns in the experiment and -3.9 ns and -7.9 ns in the simulation. The associated insertion losses are about 26 and 46 dB experimentally and amount to 35 dB and 54 dB in the simulation, respectively. Interestingly the 5.5 GHz mode, which has been observed in Fig. 5.5, appears again in Figs. 5.6 and 5.7. For all cases it holds a small negative group delay. This confirms the behavior expected from the discussion of the dispersion analysis at the end of section III. For the case of two pairs just a small negative group delay of -0.4 ns is found experimentally. For further understanding, one can go back to the basic definition of the group delay [117]:

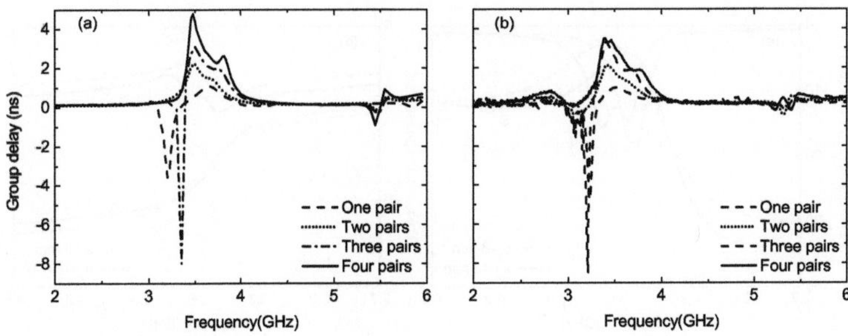

Figure 5.7: Simulated (a) and measured (b) group delay for one, two, three, and four SRRs-strip lines pairs (planar CPW structures).

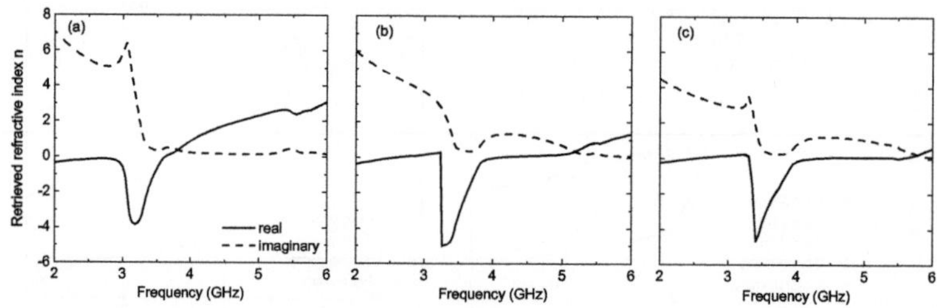

Figure 5.8: Refractive index for the one-pair (a), the two-pair (b), and three-pairs (c) structures.

$$\tau_g = \frac{L}{v_g}, v_g = \frac{c}{n_g} = \frac{c}{Re(n) + w\partial Re(n)/\partial w} \tag{5.3}$$

where v_g and n_g are the group velocity and the group index, respectively. So, if the denominator of the group velocity in Eq. (3) is negative, this yields a negative group delay. Fig. 5.8 shows the retrieved real and imaginary part of the complex refractive index using the robust method used in [118] for one-pair (a), two-pairs (b), and three-pairs (c).

It explains the behavior of the group delay. In the case of a single pair (a), a negative slope of the real part of the refractive index is noticed in the narrow frequency band between 3 GHz and 3.2 GHz. This leads to a negative group delay. For the two-pair structure, we observe an instantaneous drop from a positive level to a negative value of the real part of the refractive index, i.e. there is no frequency band with a negative slope. Hence, this behavior leads to a solely positive group delay. The little negative group delay observed experimentally for the two-pair structure can be attributed to slight tolerances in the dimensions between the real and

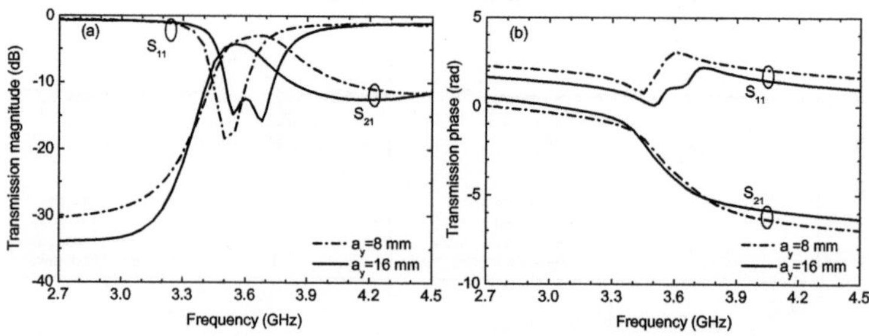

Figure 5.9: Magnitude (a) and phase (b) response for lattice constants of 8 mm and 16 mm in y-direction.

simulated structures leading to a non-abrupt drop in the refractive index. For three pairs there is again a narrow band with a negative slope. However, it is even narrower and steeper than for one pair.

In addition, the interaction between neighboring pairs might also lead to the observed difference in the group delay behavior between a single pair and multiple pairs. A simulation of the two-pair structure with a lattice constant of 16 mm in y-direction instead of 8 mm has been performed to investigate this point. Fig. 5.9 compares the transmission amplitude (a) and phase (b) of the two cases. Obviously, there is a little modification in the response, but still no negative group delay is observable even with a 16 mm distance between the two pairs. Hence, we can rule out coupling as being responsible for the absence of negative group delay for even pairs. Furthermore, we simulated only a single slot line (i.e. only one half of the CPW structure) and observed the same behavior. Therefore, the non-existence of a negative group delay cannot be attributed to mutual coupling of the two SRR-SLs belonging to one pair in conjunction with the two slots of the CPW and its modes with even/odd field symmetry. Further simulations have been carried out to investigate the effect of the number of modes excited in the CPW. All higher modes are very weak and do not show any effect related to the group delay in the resonance band. Since the ground planes on both sides of the CPW are in contact and thus at the same potential, the excitation of the parasitic mode (odd mode) is suppressed. Thus, only the even-mode propagation of the CPW exists here. In addition, simulations have been done for only the outer or only the inner ring of the SRR, for the CPW and single slot line cases. All these simulations have shown consistently only a positive group delay for an even number of the SRR-SL pairs (we studied up to eight pairs). On the other hand, negative group delay has been observed for all the structures with an odd number of pairs (we studied up to seven pairs).

5.6 Summary

We examined the behavior of the wave propagation in CPW LHM by experiments and full-wave simulations. The group delay of four coplanar waveguide (CPW) structures loaded by pairs of split ring resonators (SRRs) and strip lines (SLs) was investigated. For the first and third structures, which comprise one and three pairs of SRRs/SLs respectively, full wave simulations predict a maximum negative group delay of -3.9 ns and -7.9 ns, and a bandwidth of 160 MHz and 150 MHz, respectively. Experimentally we find a group delay of -4.7 ns and -8.5 ns and a bandwidth of 190 MHz and 290 MHz, respectively. The two-pair structure exhibits experimentally a maximum negative group delay of -0.4 ns in a 240 MHz wide band. The four structures show a positive successive growth in the group delay in the pass band as the number of SRR/SL pairs increases. In contrast to the straightforward expectation that the negative group delay can be (negatively) increased by adding more SRRs/SLs, we have shown that the opposite behavior can occur depending on the number of pairs. As a final note, since in the odd number of pairs the phase and group delay can simultaneously be made negative, these materials may be used to control the dispersive effects for an assortment of systems. We have also discussed the behavior of the group delay in the pass band for higher even numbers of SRR/SL pairs. These structures may be used as filters.

6 Novel Asymmetric Microwave and Terahertz Metasurfaces

6.1 Introduction

Various metamaterial resonators have been proposed and discussed in the last two chapters.
All the structures discussed so far were implemented by employing the coplanar waveguide. In
addition, the metamaterial research has also induced advances in related fields such as frequency
selective surfaces (FSSs) which are also known as metasurfaces or planar metamaterials. They
are used in microwave and optical filters [66, 67, 49, 119] or in thin-film sensing [120, 121, 122].
FSSs can be regarded as filters of electromagnetic waves. Composed of identical resonators
arranged periodically in two dimensions, they can be designed either for bandstop or bandpass
behavior. The first resonance of such structures appears for a wavelength slightly greater than
the array period [66]. As a result, the size of the whole FSS must be large enough compared
with the wavelength. Unfortunately, the resonating structural elements are strongly coupled to
free space and suffer from significant radiation losses so that the quality factor (Q-factor - that
is the resonance frequency over the 3-dB bandwidth of the resonator) of such structures is rather
low.

Sensing the interaction between electromagnetic waves and an unidentified sample substance provides important information for many applications throughout chemistry and biology
[123]. Especially, the recently proposed identification of polynucleotide sequences bound to
known-sequence DNA molecules at terahertz frequencies has invoked considerable interest as
it would prove advantageous over current DNA testing technologies, in which the unknown tar-
get DNA is tagged with fluorescent marker molecules [124, 125]. However, there is no robust
fingerprinting of proteins or DNA has been demonstrated using THz spectroscopy yet [125]. In
order to sense minute amounts of sample substance, thin-film sensors have to feature a sharp
transition in their frequency response, as the steepness of this transition is a direct measure for
the device sensitivity. Furthermore, the electric field has to be confined to the portion of the
sensor where the sample substance is deposited on. A variety of concepts has been introduced
in [126, 127, 128, 129, 130, 131, 132], but the proposed techniques still suffer from different
limitations [133].

The first section of this chapter presents rectangular asymmetric double split resonators
(aDSRs) with sharp tips, which offer a very high sensitivity at miniaturized scale [121]. While
aDSRs provide a stopband behavior, many applications require just the dual response, i.e. a
passband characteristic. Section 2 demonstrates applying Babinet's principle to aDSRs to appeal to this need. Finally, asymmetric single split rectangular resonators (aSRR) is introduced.

This resonator concept offers very high Q-factors well above 200, thus presenting an order of magnitude improvement compared to aDSRs, while even further reducing the electrical unit cell size.

6.2 Planar Asymmetric Metamaterial Resonators

Recently, Fedotov et al. demonstrated that resonators with weak free-space coupling can be obtained by crossing the symmetry of double split resonators (DSR) [72]. In such devices, the excitation of "trapped modes" leads to an extremely sharp resonance [72, 134]. Debus et al. simulated circular asymmetric DSR (aDSR) structures at terahertz frequencies confirming this behavior, and discussed their potential as unit cell for FSS thin-film sensors [120]. In general, asymmetry in metamaterials has been considered an unfavorable condition as it might disturb the left-handed behavior [135].

Yet, despite of the availability of resonators with weak free space coupling thin-film sensing with FSS devices is still a challenging task as the sample substance has to be either applied to a specific portion of each resonator in the array or the complete array area [120]. Depositing the sample substance at several places introduces a high degree of inaccuracy and covering the whole area is not practical in most cases, especially when only small amounts of the substance are available. Furthermore, the field confinement of a conventional circular aDSR is relatively weak, limiting its volumetric sensitivity.

In this section, we will present rectangular asymmetric double split resonators with sharp tips, which offer a very high sensitivity at miniaturized scale. Instead of a conventional FSS free-space setup, we employ a single mode rectangular waveguide with a single aDSR structure positioned inside. Thus, the experiment is more robust, as it is shielded against environmental influences, and highly reproducible. Furthermore, the waveguide confines the fields in a small region around the aDSR, significantly increasing the effective scattering cross-section of the resonator. Compared to circular structures, the rectangular design offers a miniaturization. Furthermore, the tips at the end of the resonator arms concentrate the field components into a small area, increasing the volumetric sensitivity of the device. The sensors presented here operate at C-band frequencies between 4 GHz and 8 GHz. Conducting a design study at these relatively low frequencies is advantageous as it allows the use of standard waveguide components and precise sample fabrication. However, the sensor concept can be easily miniaturized by linear scaling of its geometrical dimensions for sensing applications at higher frequencies, e.g. in the terahertz regime where rich spectroscopic features of biomolecules [136, 137] and explosives [138, 139] exist.

The three core aspects covered here are i) the experimental demonstration of aDSRs based thin-film sensors, ii) the miniaturization of the conventional circular aDSR design by employing a rectangular geometry and iii) the increase of the volumetric sensitivity by adding field confining tips to the end pieces of the resonator arms. Furthermore, the frequency offset of the resonance is analyzed for different overlayer positions and degrees of coverage, comparing the performance of the circular and the rectangular design, and the dependence of the overlayer induced resonance shift on the size of the covered area is evaluated.

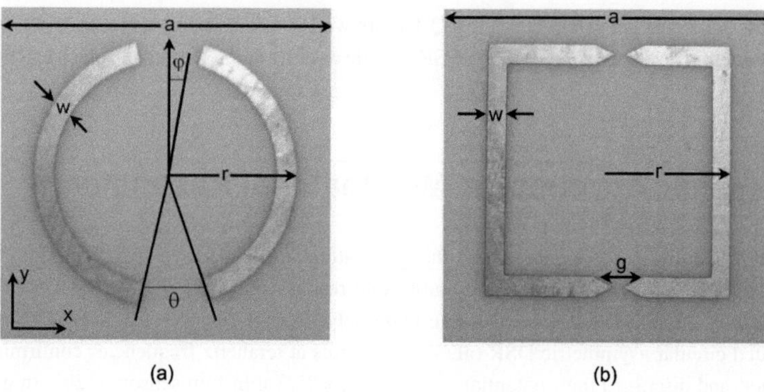

Figure 6.1: Layout of (a) the circular aDSR and (b) the rectangular aDSR with field confining
tips.

The schematic layout of the unit cells is depicted in Fig. 6.1 for both the circular and the
rectangular resonator. The dimensions are chosen such that the structure operates in the C-
band, resulting in a side length of a = 15.8 mm, an outer resonator radius of r = 4.8 mm, a
resonator width of w = 0.68 mm, a gap size of g = 1.85 mm, and a gap angle of θ = 30°.
The asymmetry angle is optimized for the maximum quality factor, which was found for ϕ
= 3°. The structures are fabricated on a 0.508 mm thick RO4003 substrate with a dielectric
constant of ϵ_r = 3.38 in a standard photolithography and wet etching process. Instead of a
conventional free-space FSS setup, we employ a single mode rectangular waveguide with one
unit cell placed inside. Apart from the high robustness and repeatability of the experiment,
utilizing a guided wave also yields a well-defined simulation environment. The measurements
employed an HP E8361A vector network analyzer and a thru-reflect-line (TRL) kit to calibrate
the WG14 rectangular waveguide. The return loss in the calibration was better than 20 dB for
the band of interest. The structures were oriented such that the x-axis marked in Fig. 6.1 was
parallel to the longer dimension of the rectangular waveguide. All simulations were conducted
with the commercial (CST) Microwave studio software [73]. Perfect electric conductor (PEC)
boundary conditions were chosen for the top and the bottom walls of the waveguide, while
periodic magnetic conductor (PMC) boundaries were applied to the sides. The general setup
has been shown in Fig. 3.11(a) which illustrates the orientation of the samples with regards to
the incoming wave. The aDSR is placed in the center of the waveguide with the plane containing
the resonator structures perpendicular to the wave vector.

The simulated (dashed lines) and measured (solid lines) reflection and transmission responses of both structures are shown in Fig. 6.2(a) and Fig. 6.2(b), respectively. A very narrow
stopband is observed around 7.7 GHz in case of the circular resonator (CR) which is shifted to
6 GHz in case of the rectangular resonator (RR). This corresponds to a miniaturization of 23%.
The reflection peak amplitude is also improved from -3.85 in case of the CR to -1.7 dB exhibited
by the RR. The reflectivity losses at the resonance frequency exceed -10 dB for both structures.
Furthermore, the transmission at the resonance frequency of the RR is reduced to -11.7 dB, re-

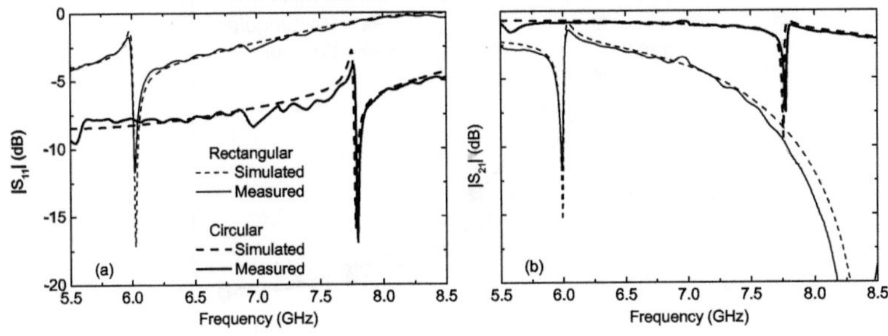

Figure 6.2: Reflection (a) and transmission (b) simulations (dashed lines) and measurements (solid lines) for CR and RR structures.

sulting in a much more pronounced stopband compared to the CR where the transmission at the resonance is found to be -7.4 dB. Moreover, the flank steepness near the resonance is improved: For the CR, 0.045 GHz lie between the minimum of -7.4 dB and the maximum of -1.1 dB. This results in a flank steepness of 140 dB/GHz. For the RR, the frequency span between the minimum of -11.7 dB and the maximum of -1.4 dB is 0.055 GHz, yielding a steepness of almost 190 dB/GHz. Especially for FSS and thin-film sensor designs, the flank steepness is considered a very important parameter [140].

To better understand the role of the tips at the end of the RR arms, the spatial distributions were calculated for both cases. Fig. 6.3(a) and Fig. 6.3(b) illustrate the spatial distribution of the absolute value of the electric field in the CR and the RR with tips, respectively. In the case of the CR, the strongest field amplitude is located at the ends of the longer resonator arm with peak values of 13.7 V/m, while the field components inside the gap remain relatively small. In contrast to that behavior, the RR with tips strongly confines the field into the gap with peak values of 17.1 V/m. This enables a high sensitivity of the proposed structure. Fig. 6.4 shows

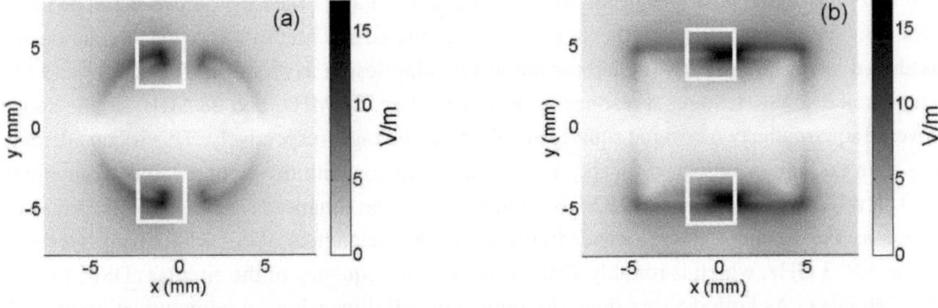

Figure 6.3: Simulated spatial field distribution in case of an excitation field strength of 1 V/m for (a) the circular aDSR and (b) the rectangular aDSR with field confining tips.

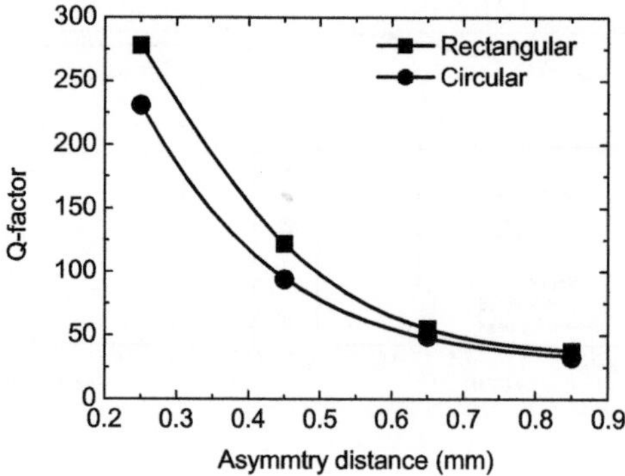

Figure 6.4: Q-factor vs. asymmetric distance for the circular aDSR and rectangular aDSR.

the Q-factor of the CR and the RR resonators versus a sweep of the asymmetry distance, which is the distance between the middle of the gap to the middle of the unit cell. The simulations have been performed for the assumption of a copper clad metallization on a lossy substrate. The Q-factor monotonically decreases with the asymmetry distance for both resonators. It starts to saturate when the asymmetry distance reaches 0.9 mm. However, the Q-factor of the rectangular resonator is higher than the one for circular one. This is interpreted to the high field confinement as seen in Fig. 6.3(b), which is due to the tips at the ends of the resonator arms.

6.2.1 Microwave Thin-Film Sensing with Asymmetric Resonators

To further characterize the structures and evaluate their potential as thin-film sensors, the aDSRs were coated with a 17.8 μm thick layer of photoresist (AZ9260). Four degrees of coverage were investigated: The uncovered aDSRs, the aDSRs with a photoresist layer applied to the whole unit cell, the aDSRs with a single 3 mm^2 square and the aDSRs with two 3 mm^2 squares. For the two latter cases, the positions of the squares were chosen as marked by the two white boxes in Fig. 6.3. The transmission coefficient (S_{21}) was simulated and measured for the circular design as shown in Fig. 6.5(a) and (c) and for the rectangular design as shown in Fig. 6.5(b) and (d). The circular aDSR features a resonance shift of 9 MHz, 24 MHz, and 48 MHz for the single covered square, the two covered squares and the full coverage, respectively. This is considerably less than the values observed for the rectangular resonator with tips, where the corresponding shifts are 18 MHz, 36 MHz and 78 MHz, demonstrating the enhanced sensitivity of the proposed structure. Furthermore, the resonance frequency of the rectangular aDSR without any overlayer lies at 5.993 GHz, which is roughly 77% of the design frequency of the circular aDSR, located at 7.716 GHz. As both devices share the same unit cell dimensions, a miniaturization by 23% is achieved.

Another interesting aspect, which is visualized in Fig. 6.5, is the fact that the insertion

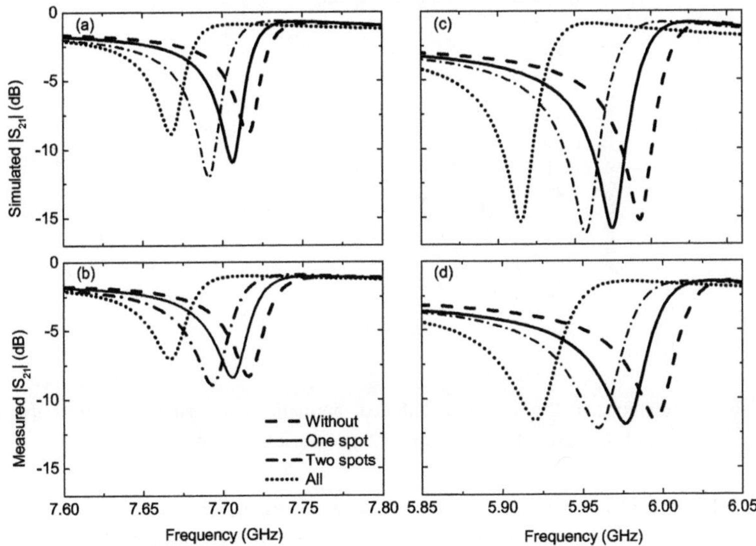

Figure 6.5: Simulated (a) and (c) and measured (b) and (d) transmission response (S_{21}) of the circular aDSR (a)(b) and rectangular aDSR (c)(d) for four degrees of coverage and an overlayer thickness of $t = 17.8\mu$m, respectively.

loss at the resonance of the aDSR does not gradually increase with the degree of coverage. In contrast, an increase is only observed in case of the single or the double square, but for the fully covered resonator, the loss is reduced again. Thus, only considering the overlayer induced dielectric losses does not suffice as explanation. Instead, the change in the asymmetry of the resonator has to be taken into account. This change in the permittivity environment increases the electrical length of the longer resonator half, leading to a larger asymmetry ratio, which results in the added insertion loss. When the whole structure is covered, the change in the permittivity environment leads to a detuning of the entire resonator, but the asymmetry ratio remains unchanged so that in this case the insertion loss only slightly differs from the uncovered case.

Fig. 6.6 illustrates the dependence of the overlayer induced frequency shift on the size of the covered area. The frequency offset, normalized to the resonance frequency of the corresponding structure, was simulated for overlayer squares with side lengths of 0.5, 1, 2, 3, 4, 5 and 6 mm. The film thickness for all simulations was set to be 17.8 μm and the squares were positioned at the end pieces of the longer resonator arm at the position of maximum field. The resulting curve progression reveals a relatively high offset sensitivity for small overlayer squares, which decreases towards larger areas. This behavior is expected from the electric field distribution

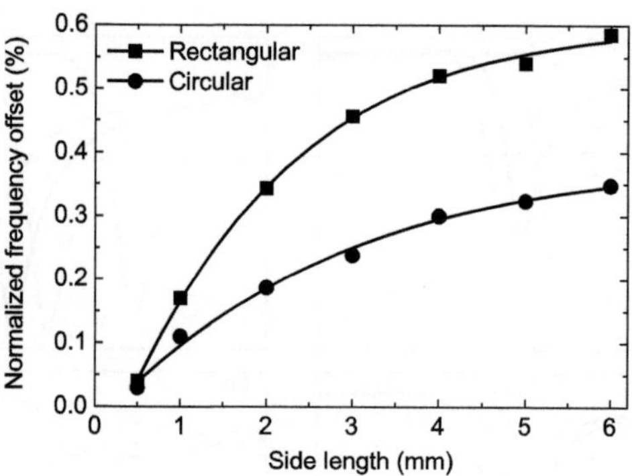

Figure 6.6: Simulations of the overlayer induced, normalized resonance shift over the side length of the covered square for a 17.8 μm thick thin-film.

shown in Fig. 6.3. Compared to the conventional circular structure, the rectangular aDSR with tips yields a stronger overlayer induced frequency shift for all simulated cases. Especially for larger areas, the offset increase reaches values of up to 90%. Hence, the tips at the end pieces of the resonator confine the field in the gap between the resonator arms, which further improves the volumetric sensitivity of the device. Next we will show how this principle can be scaled to terahertz frequencies, where rich spectroscopic features of biomaterials are present, paving the way towards highly sensitive thin-film sensors for biological and biochemical applications.

6.2.2 Terahertz Thin-Film Sensing with Asymmetric Resonators

Recently, terahertz (THz) waves have found an increasing number of applications in different fields of research [141, 142]. The THz region is located in the frequency region between microwaves and infrared. Among different applications, terahertz sensing with miniature amounts of chemical and biomolecular substances has been of particular interest to scientists and engineers alike, since it may open a broad range of potential applications. However, this is not an easy task to achieve with conventional terahertz systems given the huge difference between the sensing wavelength (sub mm) and nanometric scale of substance quantity associated with typical applications. First approaches have already been introduced as aforementioned in the introduction. However, the various techniques suffer from different limitations as mentioned earlier in the introduction.

A sharp and precise edge in the sensor frequency response is essential to enable detection of small changes in the dielectric environment. In addition, sensing small amounts of probe material needs a high concentration of the electric field distribution in a specified area. Therefore, achieving resonances with a high quality factor and low coupling to free space is essential to succeed in the field of thin-film sensing. Recently, a theoretical analysis has shown that a high-

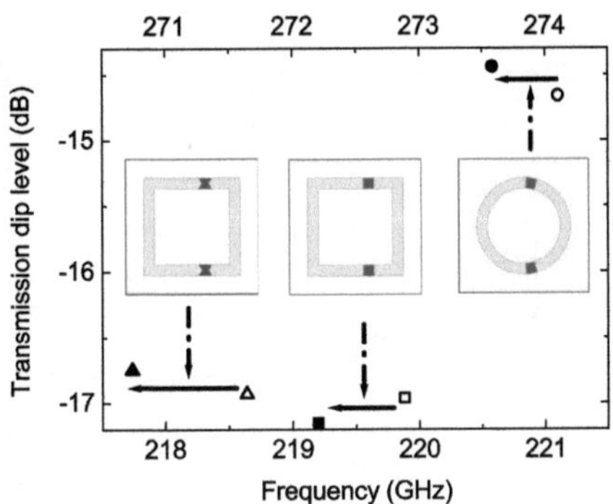

Figure 6.7: Transmission dip level vs. the resonance frequency for the three structures.

Q factor resonance can be achieved by crossing the symmetry of the double split resonators [134]. An application of this principle at microwave frequencies for different asymmetric angles of asymmetric double split resonator (aDSR) has been demonstrated experimentally [72]. The appearance of narrow resonances is ascribed to the excitation of symmetric current trapped modes, through weak free-space coupling. Simulations of such a structure have shown promising performance at terahertz frequencies as well [120]. Therefore, a miniaturized rectangular asymmetric double split resonator is proposed. Moreover, a modified design for intensified field confinement in the gaps is suggested to further increase the sensitivity for thin-film sensing. A comparative study is demonstrated to show the potential of the proposed structure [143].

We demonstrate now how a rectangular aDSR and modified rectangular aDSR with narrowing electrode shaping for increased field confinement as an improvement of the classical circular design. In order to compare the performance of the three approaches, simulations with the finite element method based on Ansoft HFSS [74] are performed. The substrate for all simulations is fused silica with a refractive index of 2.105 and a thickness of 150 μm. The lattice constant is 250 μm. The diameter of the circular structure and the rectangular side are fixed at 180 μm. The width of the gold lines is 10 μm. The gap area is fixed for all structures. The gold metallization layer and the over-layer are both 100 nm. The thin-film over-layer fills each gap completely and has the same refractive index as the substrate. In Fig. 6.7, the resonance frequency and the corresponding insertion loss, with (solid symbol) and without (hollow symbol) an over-layer, are shown for each investigated geometry. The resonance of the new rectangular aDSR is 20% lower than the one of the circular design. Thus, a miniaturization of 20% is achieved. Adding shape-confining electrodes to the rectangular resonator leads to almost doubled shift in the resonance frequency compared to the circular aDSR, improving the performance of the new design even further.

Moreover, simulations characterizing the resonance frequencies of the structures at the transmission dip as a function of over-layer thickness have carried out. These are shown for circular shape (left scale, circles), rectangular, and rectangular with tip resonators (right scale, rectangles and triangles, respectively) in Fig. 6.8. Though over-layers were applied from 25 nm to 800 nm in thickness. The shifting effect is well distinguished for the nanometer scale of

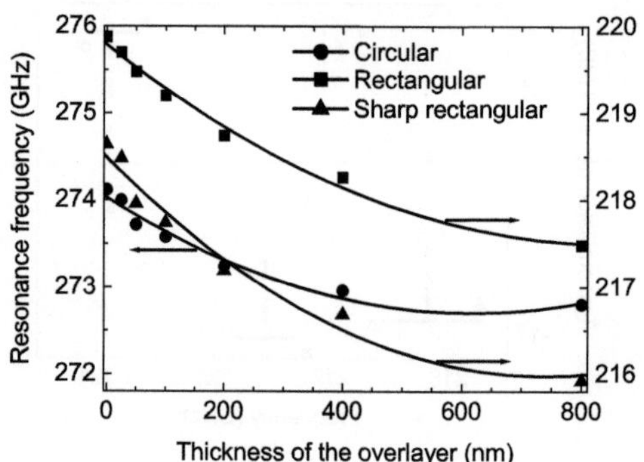

Figure 6.8: Resonance frequency at the transmission dips vs. the thickness of the over-layers (The solid lines are sigmodial fits).

the over-layer. Further increasing in the over-layer thickness shows that the saturation occurs at approximately 20 nm, and no further significant shifts were observed. Hence, the over-layer thickness of 25 nm is approximately the limit of the detect-ability of these particular structures. Therefore, these structures are very promising for future THz thin-film sensors.

6.3 Applying the Babinet Principle to Asymmetric Resonators

While aDSRs provide a stopband behavior as we have seen from the above section. However, many applications require just the dual response, i.e. a passband characteristic [49, 144, 145]. Therefore, instead of metal shaped arrays excited with electric currents, we should consider arrays of slots excited with magnetic currents. As will be shown soon, the two cases are quite comparable. It can be shown that the reflection for one array equals the transmission of its complementary. However, the metal should be a perfect conductor and infinitely thin, typically less than $\lambda/1000$ [67].

This section will present complementary asymmetric double split resonators (caDSRs), which feature a sharp passband response. The design is derived by applying the Babinet principle to asymmetric double split resonators (aDSRs) [146]. The complementary nature of these structures will be discussed in detail in regards to the transmission and reflection response as well as to the electric and magnetic field distributions.

The unit cell with a side length of $a = 15.8$ mm is shown in Fig. 6.9(a). The resonator is designed to operate in the C-band, which results in an outer radius of the metallic structures of $r = 5$ mm, a resonator width of $w = 0.8$ mm, and an opening gap angle of $\theta = 30^o$. An asymmetry angle $\phi = 10^o$ is chosen. The structures are fabricated in a standard photolithography and wet

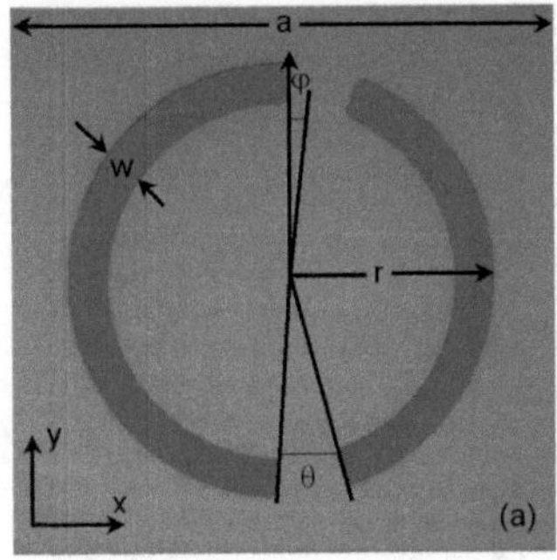

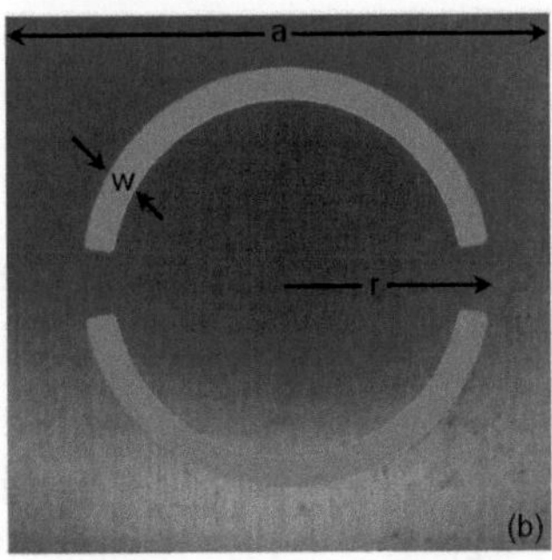

Figure 6.9: Photographs of the fabricated aDSR (a) and caDSR (b) structures with denoted dimensions.

etching process on a 0.5 mm thick FR-4 substrate with a dielectric constant of $\epsilon_r = 4.5$. We employ a single mode rectangular waveguide with a single unit cell positioned inside. Thus, the experiment is robust and highly reproducible, as it is shielded against environmental influences. Furthermore, employing a guided wave also yields a well-defined simulation environment. Interactions of an incident plane wave with the resonator can be modeled by applying the assumption that the top and bottom walls of the waveguide are perfect electric conductors (PEC). Periodic magnetic conductor (PMC) boundary conditions are assumed for the sidewalls. In case of the caDSR, Babinet's principle anticipates that each electric field component E should be interchanged with its magnetic counterpart B compared to conventional aDSR structures [147]. This implies that we need to rotate the caDSR by 90° as shown in Fig. 6.9(b). The electrical field inside the waveguide is oriented along the y-coordinate. All simulations were conducted with the commercial software Ansoft HFSS [74]. The measurements employed an HP E8361A vector network analyzer and a thru-reflect-line kit to calibrate the WG-14 waveguide. For the band of interest, the insertion and return losses in the calibration were below 0.05 dB and above 20 dB, respectively.

Fig. 6.10 show both the simulated and the measured return and insertion loss for the aDSR and the caDSR. For clarity, the y-axis in both figures is given in linear scale instead of dB. In case of the ASR, a stopband behavior with a resonance frequency at 6.53 GHz is found. The insertion loss at the resonance reaches 0.184 (14.7 dB) with corresponding return loss value of 0.84 (1.54 dB). This proves that Babinet's principle qualitatively holds true for FSSs with asymmetrical resonators, despite the fact that these structures exhibit special sharp resonances due to trapped modes. For the caDSR, the dual behavior with only a single sharp dip in the entire frequency band of interest is observed. The return loss at the resonance around 7.15 GHz is below 0.24 (12.5 dB) with a corresponding insertion loss of approximately 0.84 (1.54 dB). The quality factor reaches values of up to 27, which is high compared to other metamaterial structures, keeping in mind the presence of a lossy PCB substrate.

The discrepancy between the resonance frequency of the aDSR and the caDSR can in-part be ascribed to the effect of the dielectric substrate, which affects the resonance of the aDSR and the caDSR differently [49]. Moreover, Babinet's principle in its original formulation requires

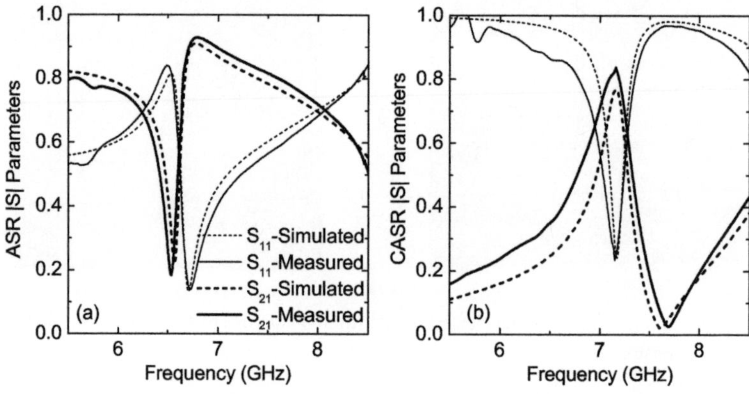

Figure 6.10: Simulated (dashed) and measured (solid) return loss (S_{11}) and insertion loss (S_{21}) of the ASR (a) and the CASR (b).

that the scattering structures are made of an infinitely thin perfect conductor, which is not the case here. The limited height of the transmission peak for the caDSR can be attributed to the presence of substrate and ohmic losses (simulations assuming the absence of the substrate and neglecting losses show a 100%-transmission). A good qualitative agreement between the simulated and measured results is achieved.

To better understand the duality of the aDSR and the caDSR, the near-field distributions were calculated for both cases. The resulting electric and magnetic field components normal to the surface at the resonance frequency of the aDSR and the caDSR are depicted in Fig. 6.11. Exactly as predicted from Babinet's principle, the magnetic near field of the caDSR exhibits an analogous behavior to the electric near field of the aDSR. Hence, the duality also applies to the electric and magnetic near-field distributions normal to the surface.

Fig. 6.12 shows the Q-factor of the caDSR versus a sweep of the asymmetry angle ϕ from 2° to 10°. The simulations have been performed for both the assumption of a perfect electric conductor metallization without any substrate (lossless case), and the assumption of a copper clad metallization on a lossy substrate (lossy case). In the lossless scenario, the Q-factor monotonically decreases with the asymmetry angle, but when losses are considered, not only the quantitative but also the qualitative behavior drastically differs. Here, the Q-factor has a clear maximum at $\phi = 5^o$, decreasing to both sides. This interesting behavior can be attributed to the fact that the trapped-mode resonance broadens and becomes weaker, as also suggested in [72] and underlines the importance of including losses in the simulations when asymmetric resonators are concerned. Therefore, the Q-factor of the caDSR structures can be custom-tailored to design requirements by varying the asymmetry angle, which opens the door for a broad range of potential applications.

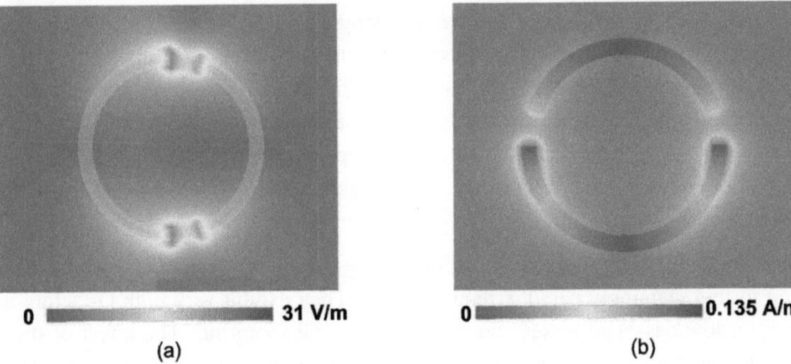

Figure 6.11: Simulated spatial field distribution of (a) the aDSR and (b) the caDSR for an excitation field amplitude of 1 V/m.

6.4 High Q-factor Metasurfaces Based on Miniaturized aSRRs

Conventional FSSs consist of periodically arranged, arbitrarily shaped metallic patches as aforementioned in the introduction. Their unit cell dimension is in the order of half the guided resonance wavelength. For some applications, e.g. low-frequency antenna radomes or frequency selective electromagnetic interference shielding, FSS unit cells of much smaller electrical size are required [148, 71]. Planar metamaterial resonators can be employed to realize such structures. Apart from the outline dimensions, the quality factor of the unit cell resonators strongly influences the resulting filter performance of the FSS. If the Q-factor of the resonators is low,

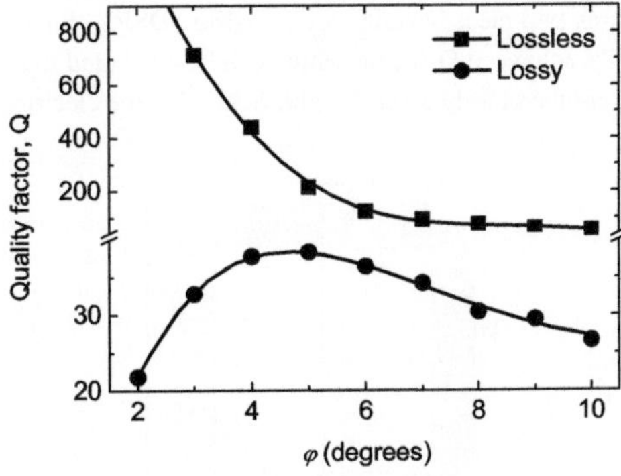

Figure 6.12: Quality factor of the caDSR versus a sweep of the asymmetry angle for the lossless and the lossy case.

cascading multiple FSS panels with quarter-wavelength spacing in between becomes necessary to still achieve a sharp transmission response leading to more complex structures with increased costs and limited applicability [122]. Consequently, metamaterial resonators combining both a small outline and a high Q-factor are highly desired, as they enable compact, high performance FSS devices. However, these design requirements appear to be contradicting, as high Q-factors require a large-volume confinement of the electromagnetic field, which can hardly be provided by sub-wavelength particles.

In the following, asymmetric single split rectangular resonators (aSRR) are introduced as a solution to this seemingly paradox requirements. They exhibit very high Q-factors in combination with low passband losses and a small electrical footprint. The effect of the degree of asymmetry on the frequency response is thoroughly studied. Furthermore, complementary structures, which feature a bandpass behavior, were derived by applying Babinet's principle and investigated with regards to their transmission characteristics. In the future, asymmetric single split rectangular resonators could provide efficient unit cells for frequency selective surface devices, such as thin-film sensors or high performance filters.

6.4.1 The Basic Principle

The working principle of the proposed asymmetric structures as depicted in Fig. 6.13 is best understood by first discussing the basic behavior of its well characterized counterpart, the symmetric single split rectangular resonator. When excited by a y-polarized electric field, the current flows equally into the right and the left arm with an out-of-phase relation. An imaginary line parallel to the y-axis defines the equilibrium point of the current distribution. When the electrical length of the resonator equals $\lambda_g/2$, the current reaches its maximum, which is equivalent to the well-known dipole resonance. From this understanding, changing the equilibrium point of the current distribution by crossing the symmetry of the structure should lead to an excitation of asymmetry resonances similar to the current trapped modes encountered in aDSRs. However, the aSRR design features two clear benefits compared to aDSRs. First, the asymmetry resonances have a drastically enhanced Q-factor, which can be attributed to the improved coupling efficiency due to the presence of only a single split. Secondly, the electrical size of the aSRR is

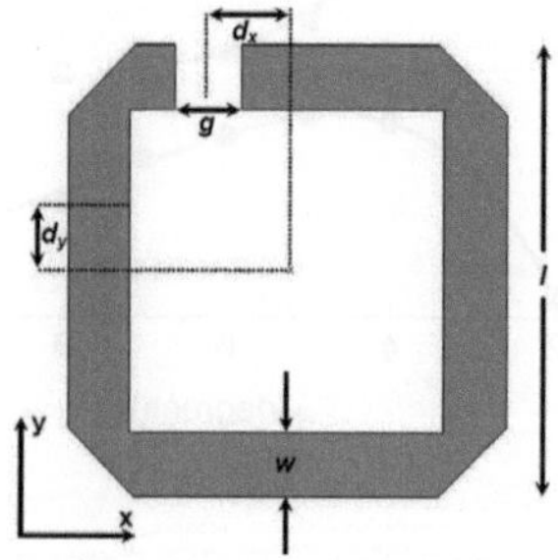

Figure 6.13: Layout of the aSRR.

less than 50% than that of the aDSR. Both, the high Q-factors and the small outline dimensions make the aSRRs very attractive for FSSs applications.

The topology of the aSRR is shown in Fig. 6.13. The dimensions are chosen such that the structure operates in the C-band. The aSRR has a side length of $l = 6.68$ mm, a width of $w = 0.8$ mm and a gap of $g = 1$ mm. The position of the gap, and thus the asymmetry of the resonator, is defined by the parameters d_x and d_y. Full symmetry is found for $d_x = 0$ and $d_y = (l - w)/2$ The structures are fabricated in a standard photolithography and wet etching process on a 0.127 mm dielectric RO3003 substrate from Rogers Corp., which has a permittivity of $\epsilon_r = 3$, and copper cladding thickness of 17 μm. To obtain a well-defined and highly reproducible measurement environment, which is not susceptible to external perturbations, a single mode rectangular waveguide is employed with the resonator unit cell centered inside. Furthermore, the guided wave conditions of the experimental setup can be easily replicated in the simulations. Here, the top and bottom walls of the waveguide are modeled as perfect electric conductors (PEC) and periodic magnetic conductor (PMC) boundary conditions are applied to the sides. All simulations were conducted with the commercial software Ansoft HFSS [74]. The measurements were again performed using an HP E8361A vector network analyzer and a thru-reflect-line kit to calibrate the WG-14 waveguide. For the band of interest, the insertion and return losses in the calibration were below 0.05 dB and above 20 dB, respectively. Please refer for Sec. 3.3.3 for more information.

First, we will continue our discussion of the basic working principle of the aSRRs by quantitatively studying the simulated S_{21} magnitude response of both a symmetric and an asymmetric SRR (Fig. 6.14). The broad dipole resonance, which is characteristic for the symmetric SRR, occurs as expected at $\lambda_g/2$, which corresponds to 15.5 GHz. The inset beside the dashed curve depicts the current distribution at the resonance frequency. As mentioned earlier, the current in both sides is equal in strength but 180° out-of-phase as marked by the black arrows. In contrary to the broad dipole like behavior of the symmetric SRR, the aSRR features a very sharp asymmetry resonance at much lower frequency. The inset near the solid curve shows the current

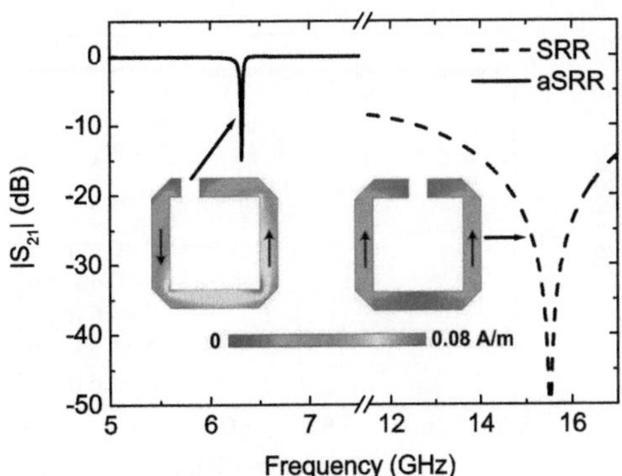

Figure 6.14: The S_{21} magnitude response of a symmetric and an asymmetric SRR vs. the frequency.

distribution at the asymmetry resonance. Three interesting features can be noticed: Firstly, the equilibrium point of the field has changed compared to the symmetric case. Secondly, the peak current is three times larger than in the symmetric case. Finally, the two parts of the current flows are in phase now so that the electric field will concentrate in a small area around the gap. This behavior is responsible for the strongly enhanced Q-factor of the aSRR compared to the symmetric SRR.

6.4.2 Parametric Study of aSRRs and caSRRs

After having discussed the operation principle of the aSRRs, we shall now study the dependence of the S_{21} magnitude of the aSRRs on the position of the split and thus on the degree of asymmetry. Six different aSRRs were simulated as shown in Fig. 6.15. For the symmetric case (d_x=0) no resonance is present. Increasing d_x leads to a more pronounced resonant behavior induced by the higher asymmetry in the current distribution and to a resonance frequency shift toward lower frequencies due to a change in the charge distribution, which leads to a modified effective gap capacitance.

While the aSRRs already constitute a significant improvement for bandstop metasurfaces, some applications require a bandpass instead of a bandstop behavior. Thus, Babinet's principle is applied to the aSRRs to derive the complementary structure, the caSRR [146]. Figs. 6.16(a) and (b) show both the simulated and the measured return and insertion losses for the aSRR and the caSRR, respectively. Measurements and simulations agree well. The small deviations can be ascribed to inaccuracies in the fabrication, especially in the wet etching process. In case of the aSRR, stopband behavior with a resonance frequency at 6.31 GHz is observed. The insertion loss at the resonance reaches 14.6 dB with a corresponding return loss value of 1.88 dB.

As expected, in case of the caSRRs, the complementary response is observed with a sharp passband around 6.5 GHz. The return loss at the resonance is below 2.9 dB with a corresponding return loss of less than 10 dB. The return loss can be fully ascribed to ohmic losses. Simulations

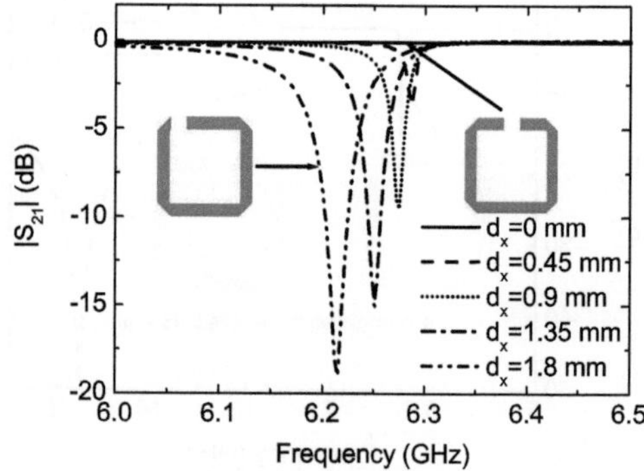

Figure 6.15: The S_{21} magnitude response of an aSRR for a sweep of the asymmetry parameter d_x.

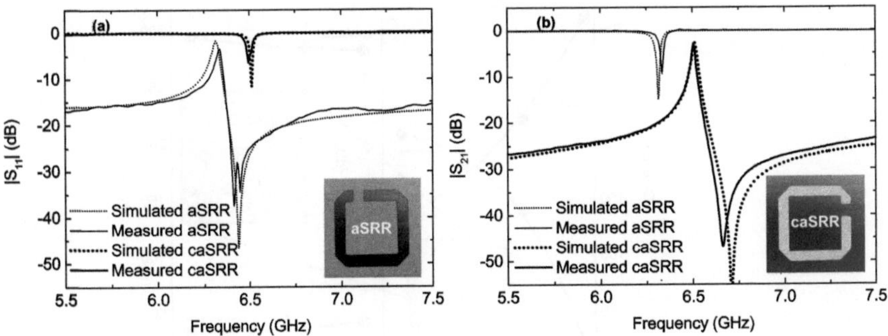

Figure 6.16: Simulated (dotted) and measured (solid) return loss (a) and insertion loss (b) of the aSRR and the caSRR.

with a lossless substrate exhibited a 100% transmission at the resonance. The 3-dB bandwidth is 29.5 MHz, despite the presence of a lossy substrate. Thus, a Q-factor of 221 is achieved, which, together with the small electrical size of just $\lambda_g/7$, clearly sets the aSRRs apart from other metamaterial resonators and underlines its potential as unit cell for high performance FSS device.

One aspect that should be briefly addressed is the discrepancy between the resonance frequency of the aSRR and the caSRR. Its origin mainly lies in the effect of the dielectric substrate, which affects the resonance of the aSRR and the caSRR differently [49]. Moreover, Babinet's principle in its original formulation [67], which is employed to derive the caSRR, requires that the scattering structures are made of an infinitely thin perfect conductor (e.g. $< \lambda/1000$), which is not fulfilled by our 17 μm thick metallization.

To illuminate the trade-off between Q-factor and insertion loss, we simulate a sweep of the split position. The simulations have been performed assuming a copper clad metallization on a lossy substrate. The gap is successively moved from a close-to-symmetry position in the right resonator arm upward, continuing in the upper resonator arm. This is equivalent to first varying d_y keeping d_x at 2.94 mm and then sweeping d_x while d_y remains at 5.88 mm. The inset in Fig. 6.17 schematically marks the eight simulated split locations. The Q-factor (outer scales) and $|S_{21}|$ insertion loss at the resonance (center scale) of the caSRR versus a sweep of d_x and d_y from 0.9 mm to 1.8 mm with a step size of 0.3 mm is shown. As expected, the losses increase with the Q-factor reaching moderate values of up to 4.8 dB for Q-factors slightly above 300. When the asymmetry in the structure is decreased, the losses reduce to 2.1 dB or 0.4 dB for Q-factors of 175 and 35, respectively. Keeping in mind that the simulations considered both the substrate and the ohmic losses the overall performance of the caSRRs is very attractive for a broad range of applications. Furthermore, the ability to select a Q-factor/insertion loss pair by only varying a single geometrical parameter, namely the split position, enables a high degree of design flexibility.

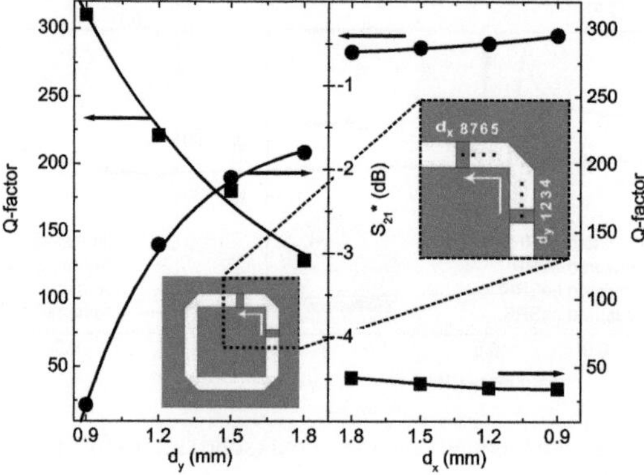

Figure 6.17: The Q-factor (left and right scales) and the insertion loss at the resonance frequency $S_{21}{}^*$ (middle scale) vs. a sweep of d_x and d_y.

6.5 Summary

In conclusion, we proposed an optimized design concept for thin-film sensors, which consists of three individual improvements. Next, we have shown that Babinet's principle qualitatively holds true for FSSs with asymmetrical resonators, despite the fact that these structures exhibit special sharp resonances due to trapped modes. Finally, we have demonstrated asymmetric single split rectangular resonators as planar metamaterials with very high Q-factors, low losses and small electrical size. Both, a bandstop and a bandpass behavior have been demonstrated. The following list summarizes the main characteristics of the previously discussed concepts.

- Rectangular Asymmetric Metamaterial Resonators (aDSR) offers many advantages over the conventional circular structures (Section 6.2, see also [121])

  - High Q-factor

  - Miniaturization of more than 20%

  - Reflection peak amplitude enhancement

  - Transmission dip level

  - Flank steepness

  - The tips at the end of the arms reveals high field confinement

- Applying the Babinet Principle to Asymmetric Resonators (Section 6.3, see also [146])

  - The Babinet principle holds true for FSSs with asymmetrical resonators

  - The duality also applies to the electric and magnetic near-field distributions normal to the surface

  - The Q-factor of the structures can be custom-tailored to design requirements

- High Q-factor Metasurfaces based on Miniaturized ASRs (Section 6.4, see also [149])

  - Very high Q-factors
  - Small electrical size-just half the aDSRs
  - High stopband rejection and low passband losses
  - Low passband losses and high stopband rejection in the case of complementary strucutres

7 Conclusions

The main objective of this dissertation was to address the challenges of simple, high performance, and miniaturized microwave circuits as well as terahertz components for filtering and sensing applications. The special properties of metamaterials are utilized to propose various kinds of resonators for different components of the system. Yet, the physical dimensions of the filters and any other resonant component are limited by the resonators' size. The principal dimension of the resonant structures needs to be at least half a wavelength. This imposes a size limitation. Correspondingly, the size of these components critically influences their performance and determine the Ohmic losses. In response to the above mentioned requirements, novel concepts for simplifying planar miniaturized metamaterials resonators for filter and sensing applications have been proposed.

A coplanar waveguide stopband structure with extraordinary electromagnetic properties based on the Babinet principle has been suggested. The resulting layout is a compact structure with a high rejecting stopband consisting of complementary split ring resonators (CSRRs) and requiring only a single metallized layer. However, the stopband response suffers from a spurious rejection band close to the main resonance, which has its origin from the difference in the electrical length of the inner and the outer resonator arm. To resolve this issue, the u-shaped CSRR is proposed, which cancels the spurious resonance by equalizing the electrical length of both resonator arms. To further miniaturize the CSRRs, rectangular multiple turn complementary spiral resonators are introduced, which enable extremely small electrical footprint filters as each turn elongates the effective resonator length, thus lowering the resonance frequency. A very small electrical size of $\lambda_g/50$ is achieved.

Later on, the effect of slots inserted into the top layer of a CPW and in direct proximity to split ring resonators is investigated. The slots enable a full control of the bandwidth by varying their lengths. This concept could also be applied to other resonator designs. Another concept to achieve very compact BPFs is by combining conventional SRRs with CSRRs. In addition to the circular structures, a rectangular spiral design (RSRs) has been investigated, which exhibits an even higher miniaturization potential, which could be very useful, e.g. in front-end filter designs.

Furthermore, the behavior of the wave propagation in coplanar waveguide left-handed media was examined. The group delay of coplanar waveguide structures loaded with pairs of split ring resonators (SRRs) and strip lines (SLs) was investigated. In contrast to the straightforward expectation that the negative group delay can be (negatively) increased by adding more SRRs/SLs, it was demonstrated that the opposite behavior can occur depending on the number of pairs. In fact, it was shown that in the odd number of pairs the phase and group delay can simultaneously be made negative, these materials may be used to control the dispersive effects in various kind of systems. Moreover, the behavior of the group delay in the pass band for higher even numbers of SRR/SL pairs was discussed. These structures may be used as filters.

Finally, a new kind of asymmetrical resonators is exploited for applications like thin-film sensing and filters. They offer a very high sensitivity at a miniaturized scale. In contrast, conventional symmetrical resonators are unable to provide the large-volume electromagnetic field confinement which is necessary to support high quality resonances. Moreover, instead of a conventional free-space setup, a single mode rectangular waveguide has been employed with a single asymmetrical double split resonator (aDSR) positioned inside. Thus, the experiment is more robust, as it is shielded against environmental influences, and highly reproducible. The above sensor concept can be easily miniaturized by linear scaling of its geometrical dimensions for sensing applications at higher frequencies, e.g. in the terahertz regime.

While aDSRs provide a stopband behavior, many applications require just the dual response, i.e. passband characteristic. I appealed to this requirement by applying Babinet's principle to aDSRs. The resulting structures are complementary asymmetric double split resonators, which offer the desired dual behavior. The quality factor of the above mentioned structures is much higher than that of conventional resonators. Yet, the electrical size still not very small. This motivated me to propose a new structure that is less than half the size of these structures. Moreover, it shows a much better performance. Both, a bandstop and a bandpass behavior have been demonstrated. Latter was achieved by applying Babinet's principle to obtain complementary structures. In the future, such metasurfaces could provide highly efficient unit cells for FSS devices.

Bibliography

[1] A. Sihvola, "Metamaterials in electromagnetics," *Metamaterials*, vol. 1, pp. 2–11, Mar. 2007.

[2] S. A. Ramakrishna, "Physics of negative refractive index materials," *Reports on Progress in Physics*, vol. 68, no. 2, pp. 449–521, 2005.

[3] V. G. Veselago, "The electrodynamics of substances with simultaneously negative values of ϵ and μ," *Soviet Physics Uspekhi*, vol. 10, no. 4, pp. 509–514, 1968.

[4] J. B. Pendry, A. J. Holden, W. J. Stewart, and I. Youngs, "Extremely low frequency plasmons in metallic mesostructures," *Physical Review Letters*, vol. 76, p. 4773, June 1996.

[5] J. B. Pendry, A. J. Holden, D. J. Robbins, and W. J. Stewart, "Magnetism from conductors and enhanced nonlinear phenomena," *IEEE Transactions on Microwave Theory and Techniques*, vol. 47, pp. 2075–2084, Nov. 1999.

[6] D. R. Smith, W. J. Padilla, D. C. Vier, S. C. Nemat-Nasser, and S. Schultz, "Composite medium with simultaneously negative permeability and permittivity," *Physical Review Letters*, vol. 84, p. 4184, May 2000.

[7] R. A. Shelby, D. R. Smith, and S. Schultz, "Experimental verification of a negative index of refraction," *Science*, vol. 292, pp. 77–79, Apr. 2001.

[8] C. Caloz, "Perspectives on EM metamaterials," *Materials Today*, vol. 12, pp. 12–20, Mar. 2009.

[9] S. A. Tretyakov, "Research on negative refraction and backward-wave media: A historical perspective," in *Negative Refraction: Revisiting Electromagnetics from Microwave to Optics*, (Lausanne, Switzerland), pp. 30–35, Feb. 2005.

[10] H. Lamb, "On group - velocity," *Proc. London Math. Soc.*, vol. s2-1, no. 1, 1904.

[11] A. Schuster, *An introduction of the theory of optics*. London: Edward Arnold, 1904.

[12] H. C. Pocklington, "Growth of a wave-group when the group-velocity is negative," *Nature*, vol. 71, pp. 607–608, 1905.

[13] L. I. Mandel'shtam, "Lecture on some problems of the theory of oscillations," *in Complete Collection of Works, Moscow: Academy of Sciences*, vol. 5, pp. 428–567, 1944.

[14] L. I. Mandel'shtam, "Group velocity in a crystal lattice," *Zh. Eksp. Teor. Fiz.*, vol. 15, pp. 475–478, 1945.

[15] J. Pendry, "Negative refraction makes a perfect lens," *Physical Review Letters*, vol. 85, pp. 3966–3969, Oct. 2000.

[16] R. Marques, F. Martin, and M. Sorolla, *Metamaterials with Negative Parameters: Theory, Design, and Microwave Applications: Theory, Design and Microwave Applications.* John Wiley & Sons, 2008.

[17] I. V. Lindell, S. A. Tretyakov, K. I. Nikoskinen, and S. Ilvonen, "BW media-media with negative parameters, capable of supporting backward waves," *Microwave and Optical Technology Letters*, vol. 31, no. 2, pp. 129–133, 2001.

[18] J. Brown, "Artificial dielectrics having refractive indices less than unity," *Proc. IEEE*, vol. 100, no. Pt. 4, pp. 51–62, 1953.

[19] W. Rotman, "Plasma simulation by artificial dielectrics and parallel-plate media," *IRE Transactions on Antennas and Propagation*, vol. 10, no. 1, pp. 82–95, 1962.

[20] J. B. Pendry, A. J. Holden, D. J. Robbins, and W. J. Stewart, "Low frequency plasmons in thin-wire structures," *Journal of Physics: Condensed Matter*, vol. 10, no. 22, pp. 4785–4809, 1998.

[21] P. Gay-Balmaz, C. Maccio, and O. J. F. Martin, "Microwire arrays with plasmonic response at microwave frequencies," *Applied Physics Letters*, vol. 81, pp. 2896–2898, Oct. 2002.

[22] G. H. B. Thompson, "Unusual waveguide characteristics associated with the apparent negative permeability obtainable in ferrites," *Nature*, vol. 175, pp. 1135–1136, June 1955.

[23] H. Hertz, *Untersuchungen über die Ausbreitung der elektrischen Kraft; translated as Electric Waves. Being Researches on the Propagation of Electric Action with Finite Velocity through Space.* Leipzig: J. A. Barth, 1891.

[24] S. Tretyakov, *Analytical Modeling in Applied Electromagnetics.* Artech House Publishers, June 2003.

[25] G. Dewar, "Candidates for $\mu < 0$, $\epsilon < 0$ nanostructures," *International Journal of Modern Physics B*, vol. 15, no. 24-25, pp. 3258–3265, 2001.

[26] "Metamorphose 2006." http://www.metamorphose-eu.org.

[27] "Defense sciences office - MetaMaterials." http://www.darpa.mil/dso/archives/metamat/.

[28] J. B. Pendry and D. R. Smith, "The quest for the superlens," *Scientific American*, vol. 295, pp. 60–7, July 2006.

[29] "Home page for professor David R. Smith." http://people.ee.duke.edu/~drsmith/.

[30] S. A. Ramakrishna and T. M. Grzegorczyk, *Physics and Applications of Negative Refractive Index Materials*. CRC, 1st ed., Sept. 2008.

[31] "World's oldest telescope?," *BBC*, July 1999.

[32] D. R. Smith, J. J. Mock, A. F. Starr, and D. Schurig, "Gradient index metamaterials," *Physical Review E*, vol. 71, p. 036609, Mar. 2005.

[33] R. B. Greegor, C. G. Parazzoli, J. A. Nielsen, M. A. Thompson, M. H. Tanielian, and D. R. Smith, "Simulation and testing of a graded negative index of refraction lens," *Applied Physics Letters*, vol. 87, no. 9, pp. 091114–1–3, 2005.

[34] T. Driscoll, D. N. Basov, A. F. Starr, P. M. Rye, S. Nemat-Nasser, D. Schurig, and D. R. Smith, "Free-space microwave focusing by a negative-index gradient lens," *Applied Physics Letters*, vol. 88, pp. 081101–1–3, Feb. 2006.

[35] J. B. Pendry, D. Schurig, and D. R. Smith, "Controlling electromagnetic fields," *Science*, vol. 312, pp. 1780–1782, June 2006.

[36] U. Leonhardt, "Optical conformal mapping," *Science*, vol. 312, pp. 1777–1780, June 2006.

[37] D. Schurig, J. J. Mock, B. J. Justice, S. A. Cummer, J. B. Pendry, A. F. Starr, and D. R. Smith, "Metamaterial electromagnetic cloak at microwave frequencies," *Science*, vol. 314, pp. 977–980, Nov. 2006.

[38] A. Lagarkov and V. Kisel, "Electrodynamic properties of simple bodies made of materials with negative permeability and negative permittivity," *Doklady Physics*, vol. 46, pp. 163–165, Mar. 2001.

[39] A. Alu and N. Engheta, "Achieving transparency with plasmonic and metamaterial coatings," *Physical Review E*, vol. 72, p. 016623, July 2005.

[40] Y. Takayama and W. Klaus, "Application of negative refractive behaviour to field-of-view expansion in optical receivers," *Journal of Optics A: Pure and Applied Optics*, vol. 5, no. 5, pp. S305–S309, 2003.

[41] N. Engheta, "An idea for thin subwavelength cavity resonators using metamaterials with negative permittivity and permeability," *IEEE Antennas and Wireless Propagation Letters*, vol. 1, pp. 10–13, 2002.

[42] G. Eleftheriades, A. Grbic, and M. Antoniades, "Negative-refractive-index transmission-line metamaterials and enabling electromagnetic applications," in *IEEE Antennas and Propagation Society International Symposium*, vol. 2, pp. 1399–1402, 2004.

[43] C. Caloz, H. Okabe, I. Awai, and T. Itoh, "Transmission line approach of left-handed materials," in *IEEE AP-S USNC/URSI National Radio Science Meeting Digest*, vol. 39, (San Antonio, TX), pp. 1159–1166, 2002.

[44] K. C. Gupta, R. Garg, I. Bahl, and P. Bhartia, *Microstrip Lines and Slotlines*. Artech House, Boston, 1996.

[45] R. N. Simons, *Coplanar Waveguide Circuits, Components and Systems*. Wiley-IEEE, New York, 2001.

[46] I. Wolff, *Coplanar Microwave Integrated Circuits*. Wiley-Interscience, July 2006.

[47] F. Falcone, F. Martin, J. Bonache, R. Marqués, and M. Sorolla, "Coplanar waveguide structures loaded with split-ring resonators," *Microwave and Optical Technology Letters*, vol. 40, no. 1, pp. 3–6, 2004.

[48] F. Martin, F. Falcone, J. Bonache, R. Marques, and M. Sorolla, "Miniaturized coplanar waveguide stop band filters based on multiple tuned split ring resonators," *IEEE Microwave and Wireless Components Letters*, vol. 13, pp. 511–513, Dec. 2003.

[49] F. Falcone, T. Lopetegi, M. A. G. Laso, J. D. Baena, J. Bonache, M. Beruete, R. Marqués, F. Martín, and M. Sorolla, "Babinet principle applied to the design of metasurfaces and metamaterials," *Phys. Rev. Lett.*, vol. 93, p. 197401, Nov. 2004.

[50] J. Garcia-Garcia, F. Martin, F. Falcone, J. Bonache, I. Gil, T. Lopetegi, M. Laso, M. Sorolla, and R. Marques, "Spurious passband suppression in microstrip coupled line band pass filters by means of split ring resonators," *IEEE Microwave and Wireless Components Letters*, vol. 14, no. 9, pp. 416–418, 2004.

[51] E. C. Jordan and K. G. Balman, *Electromagnetic Waves and Radiating Systems*. Prentice Hall Inc., 1995.

[52] P. R. Berman, "Goos-hänchen shift in negatively refractive media," *Phys. Rev. E*, vol. 66, p. 067603, Dec. 2002.

[53] M. Mojahedi and G. V. Eleftheriades, *Negative-Refraction Metamaterials: Fundamental Principles and Applications*, ch. 10. Dispersion Engineering: The Use of Abnormal Velocities and Negative Index of Refraction to Control Dispersive Effects. Wiley-IEEE Press, 2005.

[54] G. B. Arfken, H. J. Weber, and H. Weber, *Mathematical Methods for Physicists*. Academic Press, 4th ed., Sept. 1995.

[55] J. D. Jackson, *Classical Electrodynamics, 3rd ed.* John Wiley & Sons, New York, 1998.

[56] H. M. Nussenzveig, *Causality and Dispersion Relations*. Academic Press, New York, 1972.

[57] S. Ramo, J. R. Whinnery, and T. V. Duzer, *Fields and Waves in Communication Electronics*. Wiley, 3rd ed., Feb. 1994.

[58] D. R. Smith and N. Kroll, "Negative refractive index in left-handed metamaterials," *Phys. Rev. Lett.*, vol. 85, pp. 2933–2936, 2000.

[59] G. Eleftheriades, A. Iyer, and P. Kremer, "Planar negative refractive index media using periodically L-C loaded transmission lines," *IEEE Transactions on Microwave Theory and Techniques*, vol. 50, no. 12, pp. 2702–2712, 2002.

[60] M. Notomi, "Negative refraction in photonic crystals," *Optical and Quantum Electronics*, vol. 34, no. 1, pp. 133–143, 2002.

[61] R. Marques, F. Medina, and R. Rafii-El-Idrissi, "Role of bianisotropy in negative permeability and left-handed metamaterials," *Phys. Rev. B*, vol. 65, p. 144440, Apr. 2002.

[62] E. L. Bolda, R. Y. Chiao, and J. C. Garrison, "Two theorems for the group velocity in dispersive media," *Physical Review A*, vol. 48, p. 3890, Nov. 1993.

[63] E. L. Bolda, J. C. Garrison, and R. Y. Chiao, "Optical pulse propagation at negative group velocities due to a nearby gain line," *Physical Review A*, vol. 49, p. 2938, Apr. 1994.

[64] L. J. Wang, A. Kuzmich, and A. Dogariu, "Gain-assisted superluminal light propagation," *Nature*, vol. 406, pp. 277–279, July 2000.

[65] M. D. Stenner, D. J. Gauthier, and M. A. Neifeld, "The speed of information in a 'fast-light' optical medium," *Nature*, vol. 425, pp. 695–698, Oct. 2003.

[66] J. C. Vardaxoglou, *Frequency Selective Surfaces: Analysis and Design*. Research Studies Press, 1st ed., Mar. 1997.

[67] B. A. Munk, *Frequency Selective Surfaces: Theory and Design*. Wiley, New York, 2000.

[68] G. Sung, K. Sowerby, and A. Williamson, "Modeling a Low-Cost frequency selective wall for Wireless-Friendly indoor environments," *IEEE Antennas and Wireless Propagation Letters*, vol. 5, no. 1, pp. 311–314, 2006.

[69] M. Raspopoulos, P. King, and S. Stavrou, "Capacity of MIMO systems in FSS environments," *Electronics Letters*, vol. 44, no. 4, pp. 304–305, 2008.

[70] C. Mias, "Frequency selective absorption using lumped element frequency selective surfaces," *Electronics Letters*, vol. 39, no. 11, pp. 847–849, 2003.

[71] N. Qi, S. Gong, P. Zhang, and J. Liu, "A novel y-loop aperture frequency selective surface using substrate-integrated waveguide technology," *Microwave and Optical Technology Letters*, vol. 50, no. 12, pp. 3023–3027, 2008.

[72] V. A. Fedotov, M. Rose, S. L. Prosvirnin, N. Papasimakis, and N. I. Zheludev, "Sharp Trapped-Mode resonances in planar metamaterials with a broken structural symmetry," *Physical Review Letters*, vol. 99, pp. 147401–4, Oct. 2007.

[73] "CST Microwave Studio®." http://www.cst.com/Content/Products/MWS/Overview.aspx.

[74] "HFSS™." http://www.ansoft.com/products/hf/hfss.

[75] U. V. Rienen, *Numerical Methods in Computational Electrodynamics*. Springer, 2001.

[76] T. Weiland, "Time domain electromagnetic field computation with finite difference methods," *International Journal of Numerical Modelling: Electronic Networks, Devices and Fields*, vol. 9, no. 4, pp. 295–319, 1996.

[77] M. Clemens and T. Weiland, "Discrete electromagnetism with the finite integration technique," *Geometric Methods for Computational Electromagnetics, PIER*, vol. 32, pp. 65–87, 2001.

[78] *AZ 5214-E datasheet, Clariant GmbH.*

[79] J. D. Joannopoulos, *Photonic Crystals: Molding the Flow of Light*. Princeton University Press, illustrated ed., Sept. 1995.

[80] V. Radisic, Y. Qian, R. Coccioli, and T. Itoh, "Novel 2-D photonic bandgap structure for microstrip lines," *IEEE Microwave and Guided Wave Letters*, vol. 8, pp. 69–71, Feb. 1998.

[81] J. S. Hong and M. J. Lancaster, *Microstrip Filters for RF/Microwave Applications*. Wiley, 2001.

[82] R. Levy, R. Snyder, and G. Matthaei, "Design of microwave filters," *IEEE Transactions on Microwave Theory and Techniques*, vol. 50, no. 3, pp. 783–793, 2002.

[83] I. Hunter, L. Billonet, B. Jarry, and P. Guillon, "Microwave filters-applications and technology," *IEEE Transactions on Microwave Theory and Techniques*, vol. 50, no. 3, pp. 794–805, 2002.

[84] S. Liao, P. Sun, H. Chen, and X. Liao, "Compact-size coplanar waveguide bandpass filter," *IEEE Microwave and Wireless Components Letters*, vol. 13, no. 6, pp. 241–243, 2003.

[85] Y. Lin, W. Ku, C. Wang, and C. H. Chen, "Wideband coplanar-waveguide bandpass filters with good stopband rejection," *IEEE Microwave and Wireless Components Letters*, vol. 14, no. 9, pp. 422–424, 2004.

[86] M.-H. Weng, C.-Y. Hung, C.-Y. Huang, C.-S. Ye, and Y.-K. Su, "A novel compact coplanar-waveguide bandpass filter with good stopband rejection," *Microwave and Optical Technology Letters*, vol. 49, no. 2, pp. 369–371, 2007.

[87] J. Baena, J. Bonache, F. Martin, R. Sillero, F. Falcone, T. Lopetegi, M. Laso, J. Garcia-Garcia, I. Gil, M. Portillo, and M. Sorolla, "Equivalent-circuit models for split-ring resonators and complementary split-ring resonators coupled to planar transmission lines," *IEEE Transactions on Microwave Theory and Techniques*, vol. 53, no. 4, pp. 1451–1461, 2005.

[88] J. Garcia-Garcia, F. Martin, F. Falcone, J. Bonache, J. Baena, I. Gil, E. Amat, T. Lopetegi, M. Laso, J. Iturmendi, M. Sorolla, and R. Marques, "Microwave filters with improved stopband based on sub-wavelength resonators," *IEEE Transactions on Microwave Theory and Techniques*, vol. 53, no. 6, pp. 1997–2006, 2005.

[89] J. F. Woodley and M. Mojahedi, "Negative group velocity and group delay in left-handed media," *Phys. Rev. E*, vol. 70, p. 046603, Oct. 2004.

[90] I. A. Ibraheem, J. Schoebel, and M. Koch, "Group delay characteristics in coplanar waveguide left-handed media," *Journal of Applied Physics*, vol. 103, no. 2, p. 024903, 2008.

[91] J. Garcia-Garcia, F. Martin, J. D. Baena, R. Marques, and L. Jelinek, "On the resonances and polarizabilities of split ring resonators," *Journal of Applied Physics*, vol. 98, no. 3, pp. 033103–9, 2005.

[92] F. Bilotti, A. Toscano, and L. Vegni, "Design of spiral and multiple split-ring resonators for the realization of miniaturized metamaterial samples," *IEEE Transactions on Antennas and Propagation*, vol. 55, pp. 2258–2267, Aug. 2007.

[93] V. Crnojevic-Bengin, V. Radonic, and B. Jokanovic, "Left-handed microstrip lines with multiple complementary split-ring and spiral resonators," *Microwave and Optical Technology Letters*, vol. 49, no. 6, pp. 1391–1395, 2007.

[94] I. A. Ibraheem and M. Koch, "Coplanar waveguide metamaterials: The role of bandwidth modifying slots," *Applied Physics Letters*, vol. 91, no. 11, pp. 113517–1–3, 2007.

[95] H. Chen, B.-I. Wu, L. Ran, T. M. Grzegorczyk, and J. A. Kong, "Controllable left-handed metamaterial and its application to a steerable antenna," *Applied Physics Letters*, vol. 89, no. 5, pp. 053509–1–3, 2006.

[96] J. D. Baena, R. Marqués, F. Medina, and J. Martel, "Artificial magnetic metamaterial design by using spiral resonators," *Phys. Rev. B*, vol. 69, p. 014402, Jan. 2004.

[97] F. Martin, J. Bonache, F. Falcone, M. Sorolla, and R. Marques, "Split ring resonator-based left-handed coplanar waveguide," *Applied Physics Letters*, vol. 83, no. 22, pp. 4652–4654, 2003.

[98] F. Aznar, J. Bonache, and F. Martin, "Improved circuit model for left-handed lines loaded with split ring resonators," *Applied Physics Letters*, vol. 92, no. 4, pp. 043512–1–3, 2008.

[99] J. Bonache, F. Martin, F. Falcone, J. Garcia, I. Gil, T. Lopetegi, M. Laso, R. Marques, F. Medina, and M. Sorolla, "Super compact split ring resonators CPW band pass filters," in *IEEE MTT-S International Microwave Symposium Digest*, vol. 3, pp. 1483–1486, 2004.

[100] N. Dib, L. Katehi, G. Ponchak, and R. Simons, "Theoretical and experimental characterization of coplanar waveguide discontinuities for filter applications," *IEEE Transactions on Microwave Theory and Techniques*, vol. 39, no. 5, pp. 873–882, 1991.

[101] G. Ponchak and L. Katehi, "Open- and short-circuit terminated series stubs in finite-width coplanar waveguide on silicon," *IEEE Transactions on Microwave Theory and Techniques*, vol. 45, no. 6, pp. 970–976, 1997.

[102] F. Aznar, J. Garcia-Garcia, M. Gil, J. Bonache, and F. Martin, "Strategies for the miniaturization of metamaterial resonators," *Microwave and Optical Technology Letters*, vol. 50, no. 5, pp. 1263–1270, 2008.

[103] J. Bonache, F. Martin, F. Falcone, J. Garcia, I. Gil, T. Lopetegi, M. A. G. Laso, R. Marques, F. Medina, and M. Sorolla, "Compact coplanar waveguide band-pass filter at the S-band," *Microwave and Optical Technology Letters*, vol. 46, no. 1, pp. 33–35, 2005.

[104] I. A. I. Al-Naib, C. Jansen, and M. Koch, "Very compact bandpass filter based on spiral metamaterial resonators," in *34th International Conference on Infrared, Millimeter, and Terahertz Waves, Busan, Korea*, Sept. 2009.

[105] I. A. I. Al-Naib and M. Koch, "Higher degree of miniaturization with split rectangle resonators," in *38th European Microwave Conference, Amsterdam, The Netherland*, pp. 559–562, 2008.

[106] I. Al-Naib and M. Koch, "Miniaturization of coplanar waveguide bandstop filter with spiral resonators," in *2nd International Congress on Advanced Electromagnetic Materials in Microwaves and Optics, Pamplona, Spain*, pp. 815–717, 2008.

[107] I. A. I. Al-Naib, C. Jansen, and M. Koch, "Miniaturized bandpass filter based on metamaterial resonators: a conceptual study," *Journal of Physics D: Applied Physics*, vol. 41, no. 20, p. 205002, 2008.

[108] M. W. Mitchell and R. Y. Chiao, "Negative group delay and "fronts" in a causal system: An experiment with very low frequency bandpass amplifiers," *Physics Letters A*, vol. 230, pp. 133–138, June 1997.

[109] L. Brillouin, *Wave Propagation in Periodic Structures: Electric Filters and Crystal Lattices 2d Edition*. Dover Publications, 1953.

[110] S. Chu and S. Wong, "Linear pulse propagation in an absorbing medium," *Physical Review Letters*, vol. 48, p. 738, Mar. 1982.

[111] C. G. B. Garrett and D. E. McCumber, "Propagation of a gaussian light pulse through an anomalous dispersion medium," *Physical Review A*, vol. 1, p. 305, Feb. 1970.

[112] V. N. Astratov, R. M. Stevenson, I. S. Culshaw, D. M. Whittaker, M. S. Skolnick, T. F. Krauss, and R. M. de la Rue, "Heavy photon dispersions in photonic crystal waveguides," *Applied Physics Letters*, vol. 77, pp. 178–180, July 2000.

[113] S. Longhi, M. Marano, M. Belmonte, and P. Laporta, "Superluminal pulse propagation in linear and nonlinear photonic grating structures," *IEEE Journal of Selected Topics in Quantum Electronics*, vol. 9, no. 1, pp. 4–16, 2003.

[114] Z. Jian and D. M. Mittleman, "Broadband group-velocity anomaly in transmission through a terahertz photonic crystal slab," *Physical Review B*, vol. 73, pp. 115118–4, Mar. 2006.

[115] J. N. Munday and W. M. Robertson, "Negative group velocity pulse tunneling through a coaxial photonic crystal," *Applied Physics Letters*, vol. 81, no. 11, pp. 2127–2129, 2002.

[116] O. Siddiqui, S. Erickson, G. Eleftheriades, and M. Mojahedi, "Time-domain measurement of negative group delay in negative-refractive-index transmission-line metamaterials," *IEEE Transactions on Microwave Theory and Techniques*, vol. 52, no. 5, pp. 1449–1454, 2004.

[117] J. F. Woodley, M. S. Wheeler, and M. Mojahedi, "Left-handed and right-handed metamaterials composed of split ring resonators and strip wires," *Physical Review E*, vol. 71, p. 066605, June 2005.

[118] X. Chen, T. M. Grzegorczyk, B. Wu, J. Pacheco, and J. A. Kong, "Robust method to retrieve the constitutive effective parameters of metamaterials," *Physical Review E*, vol. 70, p. 016608, July 2004.

[119] R. Mittra, C. H. Chan, and T. Cwik, "Techniques for analyzing frequency selective surfaces - a review," *Proceedings of the IEEE*, vol. 76, no. 12, pp. 1593–1615, 1988.

[120] C. Debus and P. H. Bolivar, "Frequency selective surfaces for high sensitivity terahertz sensing," *Applied Physics Letters*, vol. 91, pp. 184102–1–3, Oct. 2007.

[121] I. A. I. Al-Naib, C. Jansen, and M. Koch, "Thin-film sensing with planar asymmetric metamaterial resonators," *Applied Physics Letters*, vol. 93, no. 8, pp. 083507–1–3, 2008.

[122] S. Chiam, R. Singh, J. Gu, J. Han, W. Zhang, and A. A. Bettiol, "Increased frequency shifts in high aspect ratio terahertz split ring resonators," *Applied Physics Letters*, vol. 94, pp. 064102–1–3, Feb. 2009.

[123] P. Siegel, "Terahertz technology in biology and medicine," *IEEE Transactions on Microwave Theory and Techniques*, vol. 52, no. 10, pp. 2438–2447, 2004.

[124] M. Chee, R. Yang, E. Hubbell, A. Berno, X. Huang, D. Stern, J. Winkler, D. Lockhart, M. Morris, and S. Fodor, "Accessing genetic information with high-density DNA arrays," *Science*, vol. 274, no. 5287, pp. 610–614, 1996.

[125] A. Markelz, "Terahertz dielectric sensitivity to biomolecular structure and function," *IEEE Journal on Selected Topics in Quantum Electronics*, vol. 14, no. 1, pp. 180–190, 2008.

[126] M. Nagel, P. Bolivar, B. M., H. Kurz, A. Bosserhoff, and R. Buettner, "Integrated THz technology for label-free genetic diagnostics," *Applied Physics Letters*, vol. 80, no. 1, pp. 154–1–3, 2002.

[127] B. Fischer, M. Walther, and P. Jepsen, "Far-infrared vibrational modes of DNA components studied by terahertz time-domain spectroscopy," *Physics in Medicine and Biology*, vol. 47, no. 21, pp. 3807–3814, 2002.

[128] S. Mickan, A. Menikh, H. Liu, C. Mannella, R. MacColl, D. Abbott, J. Munch, and X.-C. Zhang, "Label-free bioaffinity detection using terahertz technology," *Physics in Medicine and Biology*, vol. 47, no. 21, pp. 3789–3795, 2002.

[129] T. Baras, T. Kleine-Ostmann, and M. Koch, "On-chip THz detection of biomaterials: A numerical study," *Journal of Biological Physics*, vol. 29, no. 2-3, pp. 187–194, 2003.

[130] J. Zhang and D. Grischkowsky, "Waveguide terahertz time-domain spectroscopy of nanometer water layers," *Optics Letters*, vol. 29, no. 14, pp. 1617–1619, 2004.

[131] M. Brucherseifer, M. Nagel, P. Haring Bolivar, H. Kurz, A. Bosserhoff, and R. Buettner, "Label-free probing of the binding state of DNA by time-domain terahertz sensing," *Applied Physics Letters*, vol. 77, no. 24, pp. 4049–4051, 2000.

[132] H. Yoshida, Y. Ogawa, Y. Kawai, S. Hayashi, A. Hayashi, C. Otani, E. Kato, F. Miyamaru, and K. Kawase, "Terahertz sensing method for protein detection using a thin metallic mesh," *Applied Physics Letters*, vol. 91, no. 25, pp. 253901-1–3, 2007.

[133] J. O'Hara, R. Singh, I. Brener, E. Smirnova, J. Han, A. Taylor, and W. Zhang, "Thin-film sensing with planar terahertz metamaterials: Sensitivity and limitations," *Optics Express*, vol. 16, no. 3, pp. 1786–1795, 2008.

[134] S. Prosvirnin and S. Zouhdi, *Advances in Electromagnetics of Complex Media and Metamaterials*, ch. "Resonances of closed modes in thin arrays of complex particles", pp. 281–290. Springer Netherlands, 2003.

[135] N. Katsarakis, T. Koschny, M. Kafesaki, E. Economou, and C. Soukoulis, "Electric coupling to the magnetic resonance of split ring resonators," *Applied Physics Letters*, vol. 84, no. 15, pp. 2943–2945, 2004.

[136] B. Fischer, M. Hoffmann, H. Helm, G. Modjesch, and P. Jepsen, "Chemical recognition in terahertz time-domain spectroscopy and imaging," *Semiconductor Science and Technology*, vol. 20, no. 7, pp. S246–S253, 2005.

[137] M. Usami, M. Yamashita, K. Fukushima, C. Otani, and K. Kawase, "Terahertz wideband spectroscopic imaging based on two-dimensional electro-optic sampling technique," *Applied Physics Letters*, vol. 86, no. 14, pp. 141109-1–3, 2005.

[138] Y. Shen, T. Lo, P. Taday, B. Cole, W. Tribe, and M. Kemp, "Detection and identification of explosives using terahertz pulsed spectroscopic imaging," *Applied Physics Letters*, vol. 86, no. 24, pp. 241116-1–3, 2005.

[139] H.-B. Liu, Y. Chen, G. Bastiaans, and X.-C. Zhang, "Detection and identification of explosive RDX by THz diffuse reflection spectroscopy," *Optics Express*, vol. 14, no. 1, pp. 415–423, 2006.

[140] I. A. I. Al-Naib, C. Jansen, and M. Koch, "High degree of miniaturization with asymmetric rectangle resonators," in *3rd European Conference on Antennas and Propagation, Berlin, Germany*, March 2009.

[141] M. Tonouchi, "Cutting-edge terahertz technology," *Nat. Photon.*, vol. 1, pp. 97–105, Feb. 2007.

[142] B. Ferguson and X. Zhang, "Materials for terahertz science and technology," *Nat. Mater.*, vol. 1, no. 1, pp. 26–33, 2002.

[143] I. A. I. Al-Naib, C. Jansen, and M. Koch, "Improved terahertz sensors based on frequency selective surfaces for thin-film sensing," in *33rd International Conference on Infrared, Millimeter, and Terahertz Waves, Pasadena, California, USA*, Sept. 2008.

[144] H.-T. Chen, J. O'Hara, A. Taylor, R. Averittt, C. Highstrete, M. Lee, and W. Padilla, "Complementary planar terahertz metamaterials," *Optics Express*, vol. 15, no. 3, pp. 1084–1095, 2007.

[145] C. Rockstuhl, T. Zentgraf, T. Meyrath, H. Giessen, and F. Lederer, "Resonances in complementary metamaterials and nanoapertures," *Optics Express*, vol. 16, no. 3, pp. 2080–2090, 2008.

[146] I. Al-Naib, C. Jansen, and M. Koch, "Applying the babinet principle to asymmetric resonators," *Electronics Letters*, vol. 44, no. 21, pp. 1228–1229, 2008.

[147] M. Born and E. Wolf, *Principles of Optics*. Cambridge University Press, Cambridge, 1999.

[148] C. Mias, "Frequency selective surfaces loaded with surface-mount reactive components," *Electronics Letters*, vol. 39, no. 9, pp. 724–726, 2003.

[149] I. A. I. Al-Naib, C. Jansen, and M. Koch, "High Q-factor metasurfaces based on miniaturized asymmetric single split resonators," *Applied Physics Letters*, vol. 94, no. 15, pp. 153505-1–3, 2009.

List of Publications

Book Chapter

- I. A. I. Al-Naib, C. Jansen, and M. Koch, "Compact CPW metamaterial resonators for high performance filters", Accepted for publication in the book "Microwave and Millimeter Wave Technologies".

Journal Papers

- C. Jördens, K. L. Chee, I. A. I. Al-Naib, I. Pupeza, S. Peik, G. Wenke, and M. Koch, "Dielectric fibres for low-loss transmission of millimeter waves and its application in couplers and splitters", Journal of Infrared, Millimeter and Terahertz Waves, vol. 31, no. 2, pp. 214-220, 2010.

- I. A. I. Al-Naib and K. H. Sayidmarie, "Prediction of channel parameters from amplitude-only data using Prony algorithm", *Electronics Letters*, vol. 45, no. 15, July 2009.

- I. A. I. Al-Naib, C. Jansen, and M. Koch, "High Q-factor metasurfaces based on miniaturized asymmetric single split resonators", *Applied Physics Letters*, 94, 153505, April 2009.

- I. A. I. Al-Naib, C. Jansen, and M. Koch, "Applying the Babinet principle to asymmetric resonators", *Electronics Letters*, vol. 44, no. 21, Oct. 2008.

- I. A. I. Al-Naib, C. Jansen, and M. Koch, "Miniaturized bandpass filter based on metamaterial resonators-A conceptual study", *Journal of Physics D: Applied Physics*, 41, 205002, 2008.

- I. A. I. Al-Naib, C. Jansen, and M. Koch, "Thin-film sensing with planar asymmetric metamaterial resonators", *Applied Physics Letters*, 93, 083507, 2008.

- I. A. Ibraheem, J. Schoebel, and M. Koch, "Group delay characteristics in coplanar waveguide left-handed media", *Journal of Applied Physics*, vol. 103, 024903, Jan. 2008.

- I. A. Ibraheem, N. Krumbholz, D. Mittleman, and M. Koch, "Low-dispersive dielectric mirrors for future wireless terahertz communication systems", *IEEE Microwave and wireless components letters*, vol. 18, no. 1, Jan. 2008.

- I. A. Ibraheem and M. Koch, "Coplanar waveguide metamaterials: The role of bandwidth modifying slots", *Applied Physics Letters*, 91, 113517, 2007.

Conference Contributions

- I. A. I. Al-Naib, C. Jansen, and M. Koch, "Asymmetric single split resonators for high Q-factor metasurfaces", *34th International Conference on Infrared, Millimeter, and Terahertz Waves*, Busan, Korea, Sept. 21-25, 2009.

- I. A. I. Al-Naib, C. Jansen, and M. Koch, "Very compact bandpass filter based on spiral metamaterial resonators", *34th International Conference on Infrared, Millimeter, and Terahertz Waves*, Busan, Korea, Sept. 21-25, 2009.

- I. A. I. Al-Naib, C. Jansen, and M. Koch, "Miniaturized asymmetric resonators based on Babinet principle", *34th International Conference on Infrared, Millimeter, and Terahertz Waves*, Busan, Korea, Sept. 21-25, 2009.

- I. A. I. Al-Naib, C. Jansen, and M. Koch, "High degree of miniaturization with asymmetric rectangle resonators", *3rd European Conference on Antennas and Propagation*, Berlin-Germany, March 23-27, 2009.

- I. A. I. Al-Naib and M. Koch, "Higher degree of miniaturization with split rectangle resonators", *38th European Microwave Conference*, Amsterdam-Netherlands, Oct. 27-31, 2008.

- I. A. I. Al-Naib, C. Jansen, and M. Koch, "Miniaturization of coplanar waveguide bandstop filter with spiral resonators", *2nd International Congress on Advanced Electromagnetic Materials for Microwaves and Optics*, Pamplona-Spain, Sept. 21-26, 2008.

- I. A. I. Al-Naib, C. Jansen, and M. Koch, "Thin-film sensing with frequency selective surfaces based on improved asymmetric resonators", *2nd International Congress on Advanced Electromagnetic Materials for Microwaves and Optics*, Pamplona-Spain, Sept. 21-26, 2008.

- C. Jördens, K. L. Chee, I. A. I. Al-Naib, S. Peik, G. Wenke, and M. Koch, "Dielectric fiber based splitters, couplers and endoscopes for sub-THz frequencies", *33rd International Conference on Infrared, Millimeter, and Terahertz Waves*, Pasadena, California USA, Sept. 15-19, 2008.

- I. A. I. Al-Naib, C. Jansen, and M. Koch, "Improved terahertz sensors based on frequency selective surfaces for thin-film sensing", *33rd International Conference on Infrared, Millimeter, and Terahertz Waves*, Pasadena, California USA, Sept. 15-19, 2008.

- I. A. Ibraheem, N. Krumbholz, D. Mittleman, and M. Koch, "Low-dispersive dielectric reflectors for future wireless terahertz communication systems", *15th International Conference on Terahertz Electronics*, Cardiff-UK, Sept. 2007.

- I. A. Ibraheem and M. Koch, "Coplanar waveguide incorporating complementary split ring resonators", *1st International Congress on Advanced Electromagnetic Materials for Microwaves and Optics*, Rome-Italy, October 22-26, 2007.

- I. A. Ibraheem, J. Schoebel, and M. Koch, "Group delay in coplanar waveguide left-handed media", *1st International Congress on Advanced Electromagnetic Materials for Microwaves and Optics*, Rome-Italy, October 22-26, 2007.

- I. A. Ibraheem and J. Schoebel, "Estimation of indoor geolocation using amplitude-only data", *European Conference on Wireless Technology (ECWT'07)*, Munich, Germany, October 2007.

- I. A. Ibraheem and J. Schoebel, "Time of arrival prediction for WLAN systems using Prony algorithm", *4th workshop on positioning, navigation and communication* textit(WPNC'07), Hanover-Germany, March 2007.